D1087769

Elizabeth Reidemeister

David Hilbert, 1937

Constance Reid

Hilbert

With an appreciation
of Hilbert's mathematical work
by Hermann Weyl

Springer-Verlag New York Heidelberg Berlin 1970

Fourth Printing 1983.

ISBN 0-387-04999-1 Springer-Verlag New York Heidelberg Berlin
ISBN 3-540-04999-1 Springer-Verlag Berlin Heidelberg New York

With a frontispiece and 28 illustrations

 Library of Congress Catalog Card Number 76-97989. Printed in the United States of America.

To the memory of

Otto Blumenthal
1876—1944

Foreword

David Hilbert was one of the truly great mathematicians of his time. His work and his inspiring scientific personality have profoundly influenced the development of the mathematical sciences up to the present time. His vision, his productive power and independent originality as a mathematical thinker, his versatility and breadth of interest made him a pioneer in many different mathematical fields. He was a unique personality, profoundly immersed in his work and totally dedicated to his science, a teacher and leader of the very highest order, inspiring and most generous, tireless and persistent in all of his efforts.

To me, one of the few survivors of Hilbert's inner circle, it always has appeared most desirable that a biography should be published. Considering, however, the enormous scientific scope of Hilbert's work, it seemed to me humanly impossible that a single biographer could do justice to all the aspects of Hilbert as a productive scientist and to the impact of his radiant personality. Thus, when I learned of Mrs. Reid's plan for the present book I was at first skeptical whether somebody not thoroughly familiar with mathematics could possibly write an acceptable book. Yet, when I saw the manuscript my skepticism faded, and I became more and more enthusiastic about the author's achievement. I trust that the book will fascinate not only mathematicians but everybody who is interested in the mystery of the origin of great scientists in our society.

New Rochelle,
November 23, 1969

Richard Courant

Preface

To a large extent this book has been written from memory.

I received much friendly assistance from men and women who took their doctoral degrees from Hilbert: Vera Lebedeff-Myller (1906), Robert König (1907), Andreas Speiser (1909), Richard Courant (1910), Hugo Steinhaus (1911), Paul Funk (1911), Ludwig Föppl (1912), Hellmuth Kneser (1921), Haskell Curry (1930), Arnold Schmidt (1932), Kurt Schütte (1934).

Written recollections of other former students, no longer living, were also of great assistance. I would like to acknowledge here my special debt to Otto Blumenthal (1898), who wrote the biographical sketch for Hilbert's collected works and the one for the special edition of *Naturwissenschaften* honoring Hilbert's sixtieth birthday; and to Hermann Weyl (1908) for the obituary notice for the Royal Society and the article "David Hilbert and his mathematical work," which is reprinted in this book.

Perhaps most helpful to me because they had the longest, closest acquaintance with Hilbert were Richard Courant, who was his colleague from 1919 to 1933, most of that time as head of the Mathematical Institute; and Paul Bernays, who was from 1917 to 1934 his assistant and his collaborator in his work on logic and the foundations of mathematics.

Among Hilbert's former physics assistants, Alfred Landé, Paul Ewald, Adolf Kratzer and Lothar Nordheim were most generous with their time and knowledge. I would especially like to thank Professor Ewald for his suggestions on the literary treatment of Hilbert's life.

I was also able to obtain a great deal of information about Hilbert in personal interviews with people who, although they were not his students, were close to the Göttingen circle at various times. These included Hans Lewy, Alexander Ostrowski, George Pólya, Brigitte Rellich, Carl Ludwig Siegel, Gabor Szegö, Olga Taussky-Todd, Jan van der Corput, B. L. van der Waerden, Ellen Weyl-Bär. Letters from Kurt and Elizabeth Reidemeister and from Helmut Hasse described Hilbert's last years. Alfred Tarski and Kurt Gödel, as well as Professor Bernays, answered my questions about Hilbert's work in logic and foundations.

I am grateful to Lily Rüdenberg and Ruth Buschke for their kindness in allowing me to quote from the letters which their father, Hermann Minkowski, wrote to Hilbert during the many years of their close friendship. Unfortunately, the Hilbert half of the correspondence, which was returned to Mrs. Hilbert by Mrs. Minkowski in 1933, is — as far as I have been able to determine — no longer in existence. The few quotations from Hilbert's letters to Minkowski which do appear in this book are from Blumenthal, who had the opportunity to read Hilbert's letters before he wrote the biographical sketch for the collected works.

Horst Hilbert, the son of Hilbert's cousin, supplied many details about the family. J. K. von Schroeder of the Geheimes Staatsarchiv der Stiftung Preußischer Kulturbesitz searched out vital statistics. Kin-ya Honda translated his biographical sketch of Hilbert into English for me. H. Vogt, director of the Niedersächsische Staats- und Universitätsbibliothek, made available the letters from Hilbert which are in the Klein and Hurwitz papers. Martin Kneser, the present director of the Mathematical Institute, provided me with office space and gave me access to the Hilbert papers. Ursula Drews, the secretary of the Institute, was most helpful. Irma Neumann, whose mother was the Hilbert's well-loved housekeeper for many years, shared with me the Hilbert family pictures.

Special thanks are due to my sister, Julia Robinson, who never faltered in providing assistance, advice and encouragement; to Volker Strassen, who introduced me to Göttingen and its mathematical tradition; to Ursula Lawrenz, Christa Strassen and Edith Fried, who supplemented my knowledge of German and of Germany.

It makes me very happy that the book is being published by Springer-Verlag, who had close ties with Hilbert and Göttingen and who, by taking the risks of publication, substantially contributed to the revival of German science after the first world war.

The manuscript has been read at various stages by Paul Bernays, Richard Courant, Paul Ewald, Lothar Nordheim, Julia Robinson, R. M. Robinson, Volker Strassen, Gabor Szegö, John Addison Jr., and Max Born.

After all this very generous assistance, any errors which remain are most certainly my own.

San Francisco, California
August 3, 1969

Constance Reid

Acknowledgments

Quotations from David Hilbert's 1904 talk at Heidelberg "On the foundations of logic and arithmetic," reprinted with permission of the publisher, the Harvard University Press, Cambridge, Mass., from *From Frege to Gödel*, A Source Book in Mathematical Logic, 1879–1931. Jean van Heijenoort, Editor. Copyright 1967. Translation by Beverly Woodward.

Quotations from David Hilbert's 1925 talk at Münster "On the infinite," reprinted with permission of the publisher, Prentice-Hall, Inc., Englewood Cliffs, N. J., from *Philosophy of Mathematics:* Selected Readings. Paul Benacerraf and Hilary Putnam, Editors. Copyright 1964. Translation by Ezra Putnam and Gerald J. Massey.

Quotations from David Hilbert's 1927 talk at Hamburg "On the foundations of mathematics" and Hermann Weyl's "Comments on Hilbert's second lecture on the foundations of mathematics," reprinted with permission of the publisher, the Harvard University Press, Cambridge, Mass., from *From Frege to Gödel,* A Source Book in Mathematical Logic, 1879–1931. Jean van Heijenoort, Editor. Copyright 1967. Both translations by Stefan Bauer-Mengelberg and Dagfinn Follesdal.

Contents

I

Youth

The fortuitous combination of genes that produces an unusually gifted individual was effected by Otto Hilbert and his wife Maria sometime in the spring of 1861; and on January 23, 1862, at one o'clock in the afternoon, their first child was born in Wehlau, near Königsberg, the capital of East Prussia. They named him David.

Thanks to an autobiography and family chronicle left by the founder of the Königsberg branch of the Hilbert family, something is known about David's background on the paternal side. During the seventeenth century there were Hilberts in Saxony. For the most part they were artisans and tradespeople, but frequently enough for the fact to be commented upon they took as their wives the daughters of teachers. They were Protestants, and their Biblical names seem to indicate that they were Pietists, members of a fundamentalist sect of the time which emphasized "repentance, faith as an attitude of heart, and regeneration and sanctification as experiential facts."

At the beginning of the eighteenth century, one Johann Christian Hilbert, although trained as a brass worker, became a successful lace merchant with more than a hundred people in his employ, "the most distinguished man" in the little town of Brand near Freiberg. Unfortunately, though, he died while his children were still young, and the fortune he had left was dissipated by unscrupulous guardians. Forced by necessity, his son Christian David Hilbert apprenticed himself to a barber, served as a military barber in the army of Frederick the Great, and came eventually to Königsberg. He seems to have been a man of exceptional energy and industry. He purchased a barbershop, then enrolled at the local university, studied medicine, and became the city's licensed surgeon and accoucheur. From that time on, the Hilberts were professional men, who chose as their wives the daughters of merchants. One of Christian David's many children was David Fürchtegott Leberecht (*Fear God Live Right*) Hilbert. This was David's grandfather. He was a judge and a Geheimrat, which was a title of some honor. His son Otto was David's father, a county judge at the time of David's birth. One uncle was a lawyer, another was a gymnasium director — a position equivalent to that of a high school principal but of considerably more prestige.

Not much is known about David's background on the maternal side. Karl Erdtmann was a Königsberg merchant, and his daughter Maria Therese was David's mother. She was an unusual woman — in the German way of expression "an original" — interested in philosophy and astronomy and fascinated by prime numbers.

David's birth coincided almost exactly with the birth of German nationalism. A few months before, the brother of the dead Prussian king had made the traditional pilgrimage to Königsberg and in the ancient castle church had placed on his own head the crown of the Prussians. A short time later he chose as his chief minister the Count Otto von Bismarck-Schönhausen. In the ensuing period of the wars for the unification of Germany under Prussia, David's father became a city judge and moved his family into Königsberg proper.

The East Prussian capital dated from the middle of the thirteenth century, when the knights of the teutonic order had built their castle upon the rising ground which lies behind the point where the two branches of the Pregel river meet and flow into the Baltic. In David's time the stout castle still stood — the heart of a recently modernized city of gas lights and horse-drawn trams. From the Hilbert house at No. 13 Kirchenstrasse it was only a few blocks to the river, "our gate to freedom," as the citizens of Königsberg liked to call it. Although the city was four and a half miles from the mouth of the Pregel, the sharp salt tang of the Baltic was everywhere. Gulls settled gently on the green lawns. Sea breezes fluttered the bright sails of the fishing boats. Odors of salt water and fish, of pitch and lumber, of smoke hung over the city. Boats and barges coming up the Pregel brought romantic cargoes, unloaded and reloaded in front of tall warehouses which stood at the river's edge, and went back down to the sea with *Bernstein* (amber) and a fine white claylike substance which was used in the manufacture of pipes and called *Meerschaum*. Seven great bridges, each with a distinct personality, spanned the Pregel, five of them joining its banks with an island, called the Kneiphof. Königsberg had entered the history of mathematics through these bridges. They had provided the problem, solved the century before by Euler, which lies at the foundation of what is now known as topology. The cathedral of Königsberg was on the Kneiphof, and next to it the old university and the grave of Königsberg's greatest son, Immanuel Kant.

Like all boys in Königsberg, David grew up with the words of Kant in his mind and ears. Every year on the twenty-second of April, the anniversary of the philosopher's birth, the crypt next to the cathedral was opened to the public. On these occasions David undoubtedly accompanied his philosophically inclined mother to pay his respects, saw the bust with the familiar features crowned for this special day with a fresh wreath of laurel, and spelled out the words on the wall of the crypt:

"The greatest wonders are the starry heavens above me and the moral law within me."

His mother must also have pointed out the constellations to him and introduced him to those interesting "first" numbers which, unlike the other numbers, are divisible only by themselves and 1.

Under his father, his early training stressed the Prussian virtues of punctuality, thrift and faithfulness to duty, diligence, discipline and respect for law. A judgeship in Prussia was obtained through civil service promotion. It was a comfortable, safe career for a conservative man. Judge Hilbert is reported to have been rather narrow in his point of view with strict ideas about proper behavior, a man so set in his ways that he walked the same path every day and so "rooted" in Königsberg that he left it only for his annual vacation on the Baltic.

David was to be the Hilberts' only son. When he was six, a sister was born and christened Elise.

The year that David was seven, the King himself, soon to be Kaiser, returned to Königsberg for the first time since his coronation. Here in person for the boy to see was the man "destined," in the words of the town chronicler, "to elevate his house to its greatest brilliance, his land to its greatest strength." A large crowd gathered on the wooden bridge over the castle lake to see the King, the bridge broke under the weight, and 67 persons were drowned.

The next year, Prussia challenged France. In a few months the triumphant news blazed through the East Prussian capital — the French emperor had been taken prisoner. While Bismarck and the generals prepared to lay siege to Paris, David, now eight, started to school. This was two years later than the usual starting age of six, and it indicates that he may have received his first instruction at home, probably from his mother. She was something of an invalid and is said to have spent much of her time in bed.

In the Vorschule of the royal Friedrichskolleg, David now received the preliminary instruction required for the humanistic gymnasium, which he would have to attend if he wished to become a professional man, a clergyman, or a university professor. This included reading and writing the German and Roman scripts, spelling, parts of speech and analysis of simple sentences, important Biblical stories, and simple arithmetic involving addition, subtraction, multiplication and division of small numbers.

In the autumn of 1872, when he was ready for the gymnasium proper, the Prussian army returned in triumph to Königsberg. What was to be eventually more important for David was that at the same time a Jewish family named Minkowski came to the city from Alexoten near Kovno. They had left their native Russia because of the persecution to which Jews were being subjected by the Czar's government. The father, who had been a successful merchant, had been forced to sell everything hurriedly without profit. Now, in Königsberg, he turned to a new trade – the export of white linen rags. When his children were disturbed by this change in the family fortunes, the mother explained to them that the father's new occupation was one of the noblest, since the paper for the fine books which they loved could be made only from such rags. Eventually the father was again successful, but at first things were hard. The family moved into a big old house by the railroad station, on the other side of the Pregel from the Hilberts.

In Russia, Max, the oldest Minkowski boy, had not been permitted to enroll in the gymnasium because he was a Jew; and he was never to obtain a formal education, becoming instead a partner in his father's business and, after the latter's death, the "father" of the family. Oskar, the second son, was now one of the few Jews to attend the Altstadt Gymnasium in Königsberg. Later, as a doctor and medical researcher, he discovered the relationship between the pancreas and diabetes and became well known as the "grandfather of insulin." Eight and a half year old Hermann, the third son, entered the Vorschule of the same gymnasium. According to a loving biography compiled for the family by their sister Fanny and entitled "Three Universal Geniuses," the Minkowski boys were "a sensation" in Königsberg, "not only because of their great talents but also because of their charming personalities." Little Hermann's mathematical abilities were particularly impressive; and in one class, when the teacher failed to understand a mathematical problem on the board, the students chorused, "Minkowski, help!"

There is no sister's record that anyone was particularly impressed at this time by the abilities of the Hilbert boy. He later recalled himself as dull and silly in his youth — "dammelig" was the word he used. This was perhaps an exaggeration; but, as a friend later commented, "Behind everything Hilbert said, no matter how paradoxically it was phrased, one always felt his intense, and often quite touching, desire for the truth."

The gymnasium which David's parents had chosen for him was reputed to be the best in Königsberg, a venerable private institution of learning which dated back to the early seventeenth century and counted among its graduates Kant himself. But it was an unfortunate selection. There was at that time a rare concentration of youthful scientific talent in Königsberg. At one point Max and Willy Wien, Arnold Sommerfeld and Hermann Minkowski were all simultaneously in attendance at the Altstadt Gymnasium. David, because he attended Friedrichskolleg, did not have an opportunity to become acquainted with any of these boys during his school days.

Friedrichskolleg was also, unfortunately for David, very traditional and rigid in its curriculum. The name "Gymnasium" derived from the fact that such a school was conceived as offering a mental gymnastic which would develop a boy's mind as physical exercise develops his body. To this end the study of Latin and Greek was regarded of the utmost importance. It was believed that by cultivating these languages and their literatures, the student could acquire skill in all the mental operations. The grammar would assist him in formulating his ideas; the poetry would awaken his sense of the aesthetic and cultivate his taste; the study of historical and philosophical writers would broaden his horizons and furnish a basis for "the proper conception" of the present. After the ancient languages, mathematics was traditionally most valued for strengthening the mental muscles; but at Friedrichskolleg it ran a poor second to Latin and Greek. Science was not offered.

The language classes formed by far the largest part of the curriculum. Emphasis was on obtaining a firm foundation in grammar — the study of literature had to wait upon that. There was little opportunity for independent thinking or expression. Sometimes, however, David scribbled little verses in the margins of his notebooks.

During this same period the youngest Minkowski boy, in sheet and pillowcase, was playing Othello in a family production. Curled up in the window seat of a room where only Fanny came to practice the piano, he was reading Shakespeare, Schiller and Goethe, learning almost all of the latter by heart, "so that," according to Fanny, "for the rest of his life he needed no other books except his scientific works."

David found memorization exceedingly difficult and for the most part, at Friedrichskolleg, learning was equated with remembering. The language classes particularly, according to a friend, "caused him more sorrow than joy." He was also not particularly quick at comprehending new ideas. He seemed never really able to understand anything until he had worked it through his own mind. But in spite of his difficulties in school, he never fell behind his classmates. He was industrious, and he had a clear perception of the realities of the Prussian educational system. There were no foolish gestures. Unlike Einstein, later, he did not leave the gymnasium before taking the Abitur (the examination by means of which a student qualified to enter the university).

A member of the Hilbert family said years afterwards when she was an old lady:

"All I know of Uncle David is that his whole family considered him a bit off his head. His mother wrote his school essays for him. On the other hand, he could explain mathematics problems to his teachers. Nobody really understood him at home."

Already he had found the school subject which was perfectly suited to his mind and a source of inexhaustible delight. He said later that mathematics first appealed to him because it was "bequem" – *easy, effortless*. It required no memorization. He could always figure it out again for himself. But he recognized that according to the regulations he could not go on to the university and study mathematics, become a mathematician, unless he first obtained a diploma from a gymnasium. So now he slighted this favorite subject and concentrated on "getting by" in Latin and Greek.

The days at Friedrichskolleg were always to be remembered as unhappy ones.

There was a bright time in the year, however. That was the summer vacation. Then he accompanied his family to Rauschen, a tiny fishing village a short distance from the sea. Although Rauschen was to become a very popular summer resort, it was frequented at that time by only a few people. Among these was the large family of Karl Schmidt. Like Otto Hilbert, Schmidt had been trained as a lawyer; but he was a radical Social Democrat and had chosen eventually to be a master mason and housebuilder instead. The Schmidts' fifth child, whose name was Käthe, showed already an exceptional gift in sketching the workers and sailors of Königsberg. Many years later, as the famous artist Käthe Kollwitz, she was to recall the annual trip to Rauschen much as it must have been experienced by David:

"The trip to Rauschen took some five hours, since there was no railroad at that time. We rode in a *journalière*, which was a large covered wagon with four or five rows of seats. The back seats were taken out and filled with things needed for the stay, bedding, clothing, baskets, boxes of books and cases of wine. Three and sometimes four horses were needed. The driver sat on a high seat in front. Then off we would go through the narrow streets of Königsberg, through the clanging Tragheim Gate, and then out across the whole of Samland. Shortly before we reached Sassau we would catch sight of the sea for the first time. Then we would all stand on tiptoe and shout: *The sea, the sea!* Never again would the sea . . . be to me what the Baltic Sea at Samland was. The inexpressible splendor of the sunsets seen from the high coastline; the emotion when we saw it again for the first time, ran down the sea-slope, tore off our shoes and stockings and felt once again the cool sand underfoot, the metallic slapping of the waves . . . ."

Summers at Rauschen were idyllic. It was "a children's paradise" to those who came. In September school took up again. In November the Pregel froze and did not thaw until March.

Once, recalling his boyhood, Hilbert explained, "I did not do much mathematics in school, because I knew I would do that later on." But at the time he must have felt less philosophical about the postponement. In September 1879, at the beginning of his last gymnasium year, he transferred from Friedrichskolleg to the Wilhelm Gymnasium, a state school which placed considerably more emphasis on mathematics, treating even some of the new developments in geometry.

By this time the precocious Hermann Minkowski, although two years younger, was passing David by. That spring, "by virtue of his splendid memory and rapid comprehension" (as Hilbert later reported), Minkowski completed the eight year course at the Altstadt Gymnasium in five and a half years and went on to the local university.

At the Wilhelm Gymnasium, David was much happier than he had been at Friedrichskolleg. The teachers seemed to recognize and encourage his originality, and in later years he was often to recall them with affection. Grades improved – "good's" in almost everything (German, Latin, Greek, theology and physics) and in mathematics "vorzüglich," the highest possible mark given at the time. He did so well on his written examinations that he was excused from taking the final oral examination for the leaving certificate. The evaluation which appeared on the back of the certificate rated his deportment as "exemplary," commented on his industry and "serious scientific interest," and then concluded:

"For mathematics he always showed a very lively interest and a penetrating understanding: he mastered all the material taught in the school in a very pleasing manner and was able to apply it with sureness and ingenuity."

This is the earliest recorded glimpse of the mathematician Hilbert.

II

Friends and Teachers

It was Hilbert's good fortune that the university in his native city, although far from the center of things in Berlin, was in its scientific tradition one of the most distinguished universities in Germany.

Jacobi had taught at Königsberg when, in the time of Gauss, he was considered the "second best" mathematician in Europe. Richelot, who had been Jacobi's successor, had had the distinction of discovering in the work of an unknown gymnasium teacher the genius of Karl Weierstrass. He had then persuaded the University to award Weierstrass an honorary degree and had travelled to the little town where Weierstrass taught to make the presentation in person – "We have all found our master in Herr Weierstrass!" The versatile Franz Neumann had established at Königsberg the first institute of theoretical physics at a German university and had originated the format of the seminar.

By the time that Hilbert enrolled in the autumn of 1880, Weierstrass was the most distinguished mathematician in Germany; Jacobi and the magnanimous Richelot were dead; but Franz Neumann, who was to live to be almost a hundred, was still to be seen at university gatherings and sometimes still lectured. Every student quickly learned the story of how when a great academy had attempted to set up regulations for the apportionment of scientific credit, Neumann – many of whose discoveries were never published – had said simply, "The greatest luck is the discovery of a new truth; to that, recognition can add little or nothing."

Hilbert now found the university as free as the gymnasium had been confined. Faculty members chose the subjects they wished to teach; and students, the subjects they wished to learn. There were no specified requirements, no minimum number of units, no roll call, no examinations until the taking of the degree. Many quite naturally responded to this sudden freedom by spending their first university years in the traditional occupations of the student fraternal organizations — drinking and duelling. But for the 18-year-old Hilbert the university offered something more alluring — the freedom to concentrate, at last, upon mathematics.

There was never any doubt in Hilbert's own mind about his future vocation. Although his father disapproved, he enrolled, not in law, but in mathematics, which was at that time part of the Philosophical Faculty.

He was beginning his study at a time when much of the exuberant development of mathematics in the first part of the century had been firmly pruned into shape by Weierstrass and others. The general atmosphere was self-congratulatory. It was felt that mathematics had at last reached a level of logical strictness, or rigor, which would never need to be, and indeed could not be, surpassed. At the same time, however, a professor at Halle named Georg Cantor was developing an original theory of sets in which he treated the infinite in a new and disturbing way. According to the traditional conception, the infinite was something "unlimitedly increasing." But in this work of Cantor's it was something entirely different — not increasing but being "fixed mathematically by numbers in the definite form of a completed infinite." This conception of a "completed infinite" was one to which (Cantor later wrote) he had been "logically forced, almost against my will, because in opposition to traditions which had become valued by me." It was to be the subject of the most violent and bitter controversy among mathematicians during the coming decade.

His first semester at the university, Hilbert heard lectures on integral calculus, determinant theory, and the curvature of surfaces. The second semester, following the popular custom of moving from university to university, he set out for Heidelberg, the most delightful and most romantic of all the German universities.

At Heidelberg, Hilbert attended lectures by Lazarus Fuchs, whose name was already synonymous with linear differential equations. Fuchs's lectures were very impressive but in a rather unusual way. Rarely prepared, he customarily produced on the spot what he wished to say. Thus his students, as one of them later wrote, "had the opportunity of seeing a mathematical mind of the highest order actually in operation."

The following semester Hilbert could have moved on to Berlin, where there was a constellation of scientists that included Weierstrass, Kummer, Kronecker and Helmholtz. But he was deeply attached to the city of his birth, resembling in this at least his father; and so he returned to the University of Königsberg.

During this time there was but one full professor of mathematics at Königsberg. This was Heinrich Weber, an extremely gifted and versatile man and a worthy successor to Jacobi and Richelot. He made important contributions to fields as diverse as number theory and mathematical physics. He also wrote many important books. He co-authored with Richard Dedekind a famous book on the arithmetic theory of algebraic functions of one variable, and his book on algebra and the Riemann-Weber book on methods of mathematical physics were both classics in their field.

From Weber, Hilbert heard lectures on number theory and function theory and made his first acquaintance with the theory of invariants, the most fashionable mathematical theory of the day. He carefully saved the notes of these first lectures and also all the others he heard during his university days. The hand in which they are written is boyish, there are youthful misspellings, but no doodles. Only one set of notes appears to have been extensively worked over at a later date. These are the notes he took on the number theory lectures of Weber.

The following semester — the spring of 1882 — Hilbert chose to remain again at his home university. That same spring, Hermann Minkowski returned to Königsberg from Berlin, where he had studied for the past three semesters.

Minkowski was a chubby-faced boy with a scholar's pince-nez perched rather incongruously on his still unformed nose. While he was in Berlin, he had won a monetary prize for his mathematical work and then given it up in favor of a needy classmate. But this was not known in Königsberg. (Even his family learned of the incident only much later when the brother of the classmate told them about it.) Although Minkowski was still only 17 years old, he was involved in a deep work with which he hoped to win the Grand Prix des Sciences Mathématiques of the Paris Academy.

The Academy had proposed the problem of the representation of a number as the sum of five squares. Minkowski's investigations, however, had led him far beyond the stated problem. By the time the deadline of June 1, 1882, arrived, he still had not had his work translated into French as the rules of the competition required. Nevertheless, he decided to submit it. At the last minute, at the suggestion of his oldest brother Max, he wrote a short prefatory note in which he explained that his neglect had been due to the attractions of his subject and expressed the hope that the Academy would not think "I would have given more if I had given less." As an epigraph he inscribed a line from Boileau: "Rien n'est beau que le vrai, le vrai seul est aimable."

During the year that this work was under consideration by the Academy, Minkowski attended lectures at Königsberg. In spite of his youth, he was very stimulating to the other mathematics students. Because of his experiences in Berlin he brought to the young men isolated in the East Prussian capital a sense of participation in the mathematics of the day. He was, however, extremely shy, stammered slightly, and turned a deep red whenever attention was directed at him. It does not seem likely that during this first year he developed a close relationship with any of the other mathematics students, most of whom, like Hilbert, were several years older than he was.

Then, in the spring of 1883, came the announcement that this boy, still only 18 years old, had been awarded jointly with the well-known English mathematician Henry Smith the Grand Prix des Sciences Mathématiques. The impression which the news made in Königsberg can be gauged by the fact that Judge Hilbert admonished David that presuming on acquaintance with "such a famous man" would be "impertinence."

For a while it seemed, though, that Minkowski might not actually receive his prize. The French newspapers pointed out that the rules of the competition had specifically stated that all entries must be in French. The English mathematicians let it be known that they considered it a reflection upon their distinguished countryman, who had since died, that he should be made to share a mathematical prize with a boy. ("It is curious to contemplate at a distance," an English mathematician remarked some forty years later, "the storm of indignation which convulsed the mathematical circles of England when Smith, bracketed after his death with the then unknown German mathematician, received a greater honor than any that had been paid to him in life.") In spite of the pressures upon them, the members of the prize committee never faltered. From Paris, Camille Jordan wrote to Minkowski: "Work, I pray you, to become a great mathematician."

Hilbert knew his luck when he saw it. In spite of his father's disapproval, he soon became friends with the shy, gifted Minkowski. He was shortly to comment of another shy young mathematician that "with skillful treatment I am sure he would open up," and now he apparently applied such skillful treatment to Minkowski.

Although the two young men came from different family backgrounds and had in many ways quite different personalities, they had – under the surface – many traits in common; and, years later, when Hilbert had occasion to write about Minkowski, he revealed more about himself than he ever did at any other time.

In addition to their enthusiastic love for mathematics, they shared a deep, fundamental optimism. As far as science in general was concerned, the period of their university days was one of triumphant pessimism, a reaction against the almost religious belief in the power of science which had flourished in the previous century. The works of Emil duBois-Reymond, a physiologist turned philosopher, were widely read and much quoted. DuBois-Reymond concerned himself with the limits of the knowledge of nature — this was, in fact, the title of his most famous lecture. He maintained that certain problems, which he called transcendental, or supersensible, were unsolvable even in principle. These included the nature of matter and force, the origin of motion, the origin of sensation and consciousness. His gloomy concession, "Ignoramus et ignorabimus" — *we are ignorant and we shall remain ignorant* —, was the catchword of many of the scientific-philosophical discussions at the university. But to both Hilbert and Minkowski such a concession was thoroughly abhorrent. Already they shared the conviction (as Hilbert later put it) "that every definite mathematical problem must necessarily be susceptible of an exact settlement, either in the form of an actual answer to the question asked, or by the proof of the impossibility of its solution and therefore the necessary failure of all attempts."

This belief in the solvability of every mathematical problem had recently received spectacular support. The German mathematician Ferdinand Lindemann had proved the long suspected transcendence of the number π and had thus established the impossibility of the ancient dream of "squaring the circle." Up to the time of this achievement, Lindemann's career had not been a great success. An ambitious work which he had published had been cruelly (and rather unfairly) criticized. But now, he had recouped everything by solving a famous problem. He was the man of the hour in mathematics. When Weber left Königsberg for Charlottenburg, Lindemann was invited to take his place.

In spite of his current fame, Lindemann was not a mathematician of the same caliber as Weber. He was to have little influence on Hilbert (and none at all on Minkowski); but he was responsible for bringing to Königsberg shortly the young man who was to be, rather than either Weber or Lindemann, Hilbert's real teacher. This was Adolf Hurwitz.

In the spring of 1884 when Hurwitz arrived from Göttingen to take over the duties of an Extraordinarius, or associate professor, he was still not 25 years old. Like Minkowski, he had behind him a record of mathematical precocity. His gymnasium teacher, Hannibal Schubert, had been so impressed with his mathematical abilities that he had devoted Sundays to instructing Hurwitz in his own specialty, which became known as the "Schubert calculus." He had also managed to convince Hurwitz's father, a Jewish manufacturer as doubtful of the rewards of an academic life as Judge Hilbert, that the gifted boy must be permitted to continue his study of mathematics. Encouraged by Schubert, the father borrowed the necessary money from a friend.

Hurwitz's first mathematical work was published in collaboration with Schubert while he was still in the gymnasium. His later studies gave him an exceptionally wide background in the mathematics of the day. He received his doctor's degree from Felix Klein, one of the most spectacular of the younger mathematicians in Germany at that time. He attended the lectures of the great men in Berlin and then moved on to Göttingen, where he did impressive work in function theory.

He was a sweet-tempered young man who loved music almost as much as mathematics and played the piano beautifully. But already, before he came to Königsberg, he had suffered a nearly fatal bout with typhoid fever. He had frequent, very severe migraine headaches, which may have been at least partially caused by the fact that he was a perfectionist in everything he did.

Hilbert found the new teacher "unpretentious in his outward appearance" but saw that "his wise and gay eyes testified as to his spirit." He and Minkowski soon established a close relationship with Hurwitz. Every afternoon, "precisely at five," the three met for a walk "to the apple tree." It was at this time that Hilbert found a way of learning infinitely preferable to poring over dusty books in some dark classroom or library.

"On unending walks we engrossed ourselves in the actual problems of the mathematics of the time; exchanged our newly acquired understandings, our thoughts and scientific plans; and formed a friendship for life."

Learning "in the most easy and interesting way," the three young men explored every kingdom of the mathematical world. Hurwitz with his vast "well founded and well ordered" knowledge was always the leader. He quite overwhelmed the other two.

"We did not believe," Hilbert recalled later, "that we would ever bring ourselves so far."

But there was no need for them to feel like Alexander, who complained to his schoolmates, "Father will conquer everything and there will be nothing left for us to conquer."

The world of mathematics is inexhaustible.

III

Doctor of Philosophy

Having completed the eight university semesters required for a doctor's degree, Hilbert began to consider possible subjects for his dissertation. In this work he would be expected to make some sort of original contribution to mathematics. At first he thought that he might like to investigate a generalization of continued fractions; and he went to Lindemann, who was his "Doctor-Father," with this proposal. Lindemann informed him that unfortunately such a generalization had already been given by Jacobi. Why not, Lindemann suggested, take instead a problem in the theory of algebraic invariants.

Although the theory of algebraic invariants was considered a very modern subject, its roots lay in the seventeenth century invention of analytic geometry by René Descartes. On Descartes's map of the plane, horizontal coordinates are real numbers which are designated by x; vertical coordinates, real numbers designated by y. Since any point on the plane is then equivalent to a pair of real numbers x, y, geometric figures can be formalized by algebraic equations and, conversely, algebraic equations can be graphed as geometric figures. Concepts and relations in both fields are clarified — geometric ideas becoming more abstract and easily handled; algebraic ideas, more vivid and more intuitively comprehensible.

There is also a great gain in generality. Just as the size and the shape of figures do not change when their position in relation to the axes is changed, so certain properties of their related algebraic forms remain unchanged. These "invariants" serve to characterize the given geometric figure. Thus, quite naturally, the development of projective geometry, which concerns itself with the often dramatic transformations effected by projection, led eventually to a parallel development in algebra which concentrated on the invariants of algebraic forms under various groups of transformations.

Because of its sheer sophisticated power, the algebraic approach soon outstripped the geometric one; and the theory of algebraic invariants became a subject of consuming interest for a number of mathematicians.

The pioneers in the new theory had been Englishmen — Arthur Cayley and his good friend, Joseph Sylvester, both men, as it happened, lawyers turned mathematicians. But the Germans had been quick to take up the theory; and now the great German mathematical journal, *Mathematische Annalen*, was almost exclusively an international forum for papers on algebraic invariants.

The problem which Lindemann suggested to Hilbert for his doctoral dissertation was the question of the invariant properties for certain algebraic forms. This was an appropriately difficult problem for a doctoral candidate, but not so difficult that he could not be expected to solve it. Hilbert showed his originality by following a different path from the one generally believed to lead to a solution. It was a very nice piece of work, and Lindemann was satisfied.

A copy of the dissertation was dispatched to Minkowski, who after his father's recent death had gone to Wiesbaden with his mother.

"I studied your work with great interest," Minkowski wrote to Hilbert, "and rejoiced over all the processes which the poor invariants had to pass through before they managed to disappear. I would not have supposed that such a good mathematical theorem could have been obtained in Königsberg!"

On December 11, 1884, Hilbert passed the oral examination. The next and final ordeal, on February 7, 1885, was the public promotion exercise in the Aula, the great hall of the University. At this time he had to defend two theses of his own choice against two fellow mathematics students officially appointed to be his "opponents." (One of these was Emil Wiechert, who later became a well-known seismologist.) The contest was generally no more than a mock battle, its main function being to establish that the candidate could perceive and frame important questions.

The two propositions which Hilbert chose to defend spanned the full breadth of mathematics. The first concerned the method of determining absolute electromagnetic resistance by experiment. The second pertained to philosophy and conjured up the great ghost of Immanuel Kant.

It had been the position of Kant, who had lectured on mathematics as well as philosophy when he taught at Königsberg, that man possesses certain notions which are not *a posteriori* (that is, obtained from experience) but *a priori*. As examples of *a priori* knowledge he had cited the fundamental concepts of logic, arithmetic and geometry — among these the axioms of Euclid.

The discovery of *non*-euclidean geometry in the first part of the nineteenth century had cast very serious doubt on this contention of Kant's; for it had shown that even with one of Euclid's axioms negated, it is still possible to derive a geometry as consistent as euclidean geometry. It thus became clear that the knowledge contained in Euclid's axioms was *a posteriori* — from experience — not *a priori.*

Could this also possibly be true of the fundamental concepts of arithmetic?

Gauss, who was apparently the first mathematician to be aware of the existence of non-euclidean geometries, wrote at one time:

"I am profoundly convinced that the theory of space occupies an entirely different position with regard to our knowledge *a priori* from that of [arithmetic]; that perfect conviction of the necessity and therefore the absolute truth which is characteristic of the latter is totally wanting in our knowledge of the former. We must confess, in all humility, that number is solely a product of our mind. Space, on the other hand, possesses also a reality outside our mind, the laws of which we cannot fully prescribe *a priori.*"

Hilbert appears to have felt this way too; for in his second proposition he maintained:

That the objections to Kant's theory of the a priori *nature of arithmetical judgments are unfounded.*

There is no record of his defense of this proposition. Apparently, his arguments were convincing; for at the conclusion of the disputation he was awarded the degree of Doctor of Philosophy.

The Dean administered the oath:

"I ask you solemnly whether by the given oath you undertake to promise and confirm most conscientiously that you will defend in a manly way true science, extend and embellish it, not for gain's sake or for attaining a vain shine of glory, but in order that the light of God's truth shine bright and expand."

That night the new Doctor of Philosophy and his celebrating friends wired the news to Minkowski.

Hilbert was now set upon the first step of an academic career. If he were fortunate and qualified, he would arrive ultimately at the goal, the full professorship — a position of such eminence in the Germany of the day that professors were often buried with their title and the subject of their specialty on their gravestones. As a mere doctor of philosophy, however, he was not yet eligible even to lecture to students. First he had to turn out still another piece of original mathematical research for what was known as "Habilitation." If this was acceptable to the faculty, he would then be awarded the *venia legendi*, which carried with it the title of Privatdozent and the privilege of delivering lectures without pay under the sponsorship of the university. As such a docent he would have to live on fees paid by students who chose to come to hear his lectures. Since the courses which all the students took, such as calculus, were always taught by a member of the official faculty, he would be fortunate to draw a class of five or six. It was bound to be a meager time. Eventually, however, if he attracted attention by his work and abilities (or better yet, it was rumored, if he married a professor's daughter), he would become an Extraordinarius, or associate professor, and receive a salary from the university. The next step would be an offer of an Ordinariat, or full professorship. But this final step was by no means automatic, since the system provided an almost unlimited supply of docents from which to draw a very limited number of professors. Even in Berlin, there were only three mathematics professors; at most Prussian universities, two; at Königsberg, only one.

As a hedge against the vicissitudes of such a career, a young doctor could take the state examination and qualify himself for teaching at the gymnasium level. This was not a prize to be scorned. Although many, their eyes on the prestige-ladened professorship, didn't consider the alternative, one needed only to match the number of docents with the number of professorial chairs which might reasonably become vacant in the next decade to see its advantages. Hilbert now began to prepare himself for the state examination, which he passed in May 1885.

That same summer Minkowski returned to Königsberg, received the degree of doctor of philosophy, and then left almost immediately for his year in the army. (Hilbert was one of his official opponents for the promotion exercises.)

Hilbert had not been called up for military service. He considered a study trip, and Hurwitz urged Leipzig — and Felix Klein.

Although Klein was only 36 years old, he was already a legendary figure in mathematical circles. When he had been 23 (Hilbert's present age) he had been a full professor at Erlangen. His inaugural lecture there had made mathematical history as the Erlangen Program — a bold proposal to use the group concept to classify and unify the many diverse and seemingly unrelated geometries which had developed since the beginning of the century. Early in his career he had shown an unusual combination of creative and organizational abilties and a strong drive to break down barriers between pure and applied science. His mathematical interest was all-inclusive. Geometry, number theory, group theory, invariant theory, algebra — all had been combined for his master work, the development and completion of the great Riemannian ideas on geometric function theory. The crown of this work had been his theory of automorphic functions.

But the Klein whom Hilbert now met in Leipzig in 1885 was not this same dazzling prodigy. Two years before, in the midst of the work on automorphic functions, a young mathematician from a provincial French university had begun to publish papers which showed that he was concentrating his efforts in the same direction. Klein immediately recognized the caliber of his competitor and began a nervous correspondence with him. With almost super-human effort he drove himself to reach the goal before Henri Poincaré. The final result was essentially a draw. But Klein collapsed. When Hilbert came to him, he had only recently recovered from the long year of deep mental depression and physical lassitude which had followed his breakdown. He had passed the time writing a little book on the icosahedron which was to become a classic; but the future direction of his career was still not determined.

Hilbert attended Klein's lectures and took part in a seminar. He could not have avoided being impressed. Klein was a tall, handsome, dark haired and dark bearded man with shining eyes, whose mathematical lectures were universally admired and circulated even as far as America. As for Klein's reaction to the young doctor from Königsberg — he carefully preserved the Vortrag, or lecture, which Hilbert presented to the seminar and he later said: "When I heard his Vortrag, I knew immediately that he was the coming man in mathematics."

Since his breakdown, Klein had received two offers of positions — the first, to Johns Hopkins University in America, he had refused; the second, to Göttingen, he had just recently accepted. Apparently Hilbert had already absorbed a feeling for Göttingen, the university of Gauss, Dirichlet and Riemann. Inspired by Klein's appointment, he now scribbled one of his little verses on the inside cover of a small notebook purchased in Leipzig. The writing is so illegible that the German cannot be made out exactly, but the sense of the verse is essentially "Over this gloomy November day/ Lies a glow, all shimmery/ Which Göttingen casts over us/ Like a youthful memory."

In Leipzig, Hilbert soon became acquainted with several other young mathematicians. One of these was Georg Pick, whose knowledge of pollination and breeding as well as his admiration for Hurwitz's work recommended him to Hilbert. Another was Eduard Study, whose main interest, like Hilbert's own, was invariant theory. The two should have had a lot in common, but this was not to be the case. Study was a "strange person," Hilbert wrote to Hurwitz, "and almost completely at opposite poles from my nature and, as I think I can judge, from yours too. Dr. Study approves, or rather he knows, only one field of mathematics and that's the theory of invariants, very exclusively the symbolic theory of invariants. Everything else is unmethodical 'fooling around' .... He condemns for this reason all other mathematicians; even in his own field he considers himself to be the only authority, at times attacking all the other mathematicians of the symbolic theory of invariants in the most aggressive fashion. He is one who condemns everything he doesn't know whereas, for example, my nature is such I am most impressed by just that which I don't yet know." (Hurwitz wrote back, "This personality is more repugnant to me than I can tell you; still, in the interest of the young man, I hope that you see it a little too darkly.")

There were a considerable number of people at Leipzig who were interested in invariant theory; but Klein went out of his way to urge both Study and Hilbert to go south to Erlangen to pay a visit to his friend Paul Gordan, who was universally known at that time as "the king of the invariants." For some reason the expedition was not made. Perhaps because Hilbert did not care to make it with Study.

Hilbert was soon a member of the inner mathematical circle in Leipzig. At the beginning of December 1885, a paper of his on invariants was presented by Klein to the scientific society. On New Year's Eve he was invited to a "small but very select" party at Klein's – "Professor Klein, his honored spouse, Dr. Pick and myself." That same evening Minkowski, stranded and cold at Fort Friedrichsburg in the middle of the Pregel, was sending off New Year's greetings to his friend with the plaintive question, "Oh, where are the times when this poor soldier was wont to busy himself over the beloved mathematics?" But at the Kleins' the conversation was lively – "on all possible and impossible things." Klein tried to convince Hilbert that he should go to Paris for a semester of study before he returned to Königsberg. "He said," Hilbert wrote to Hurwitz, "that Paris is at this time a beehive of scientific activity, particularly among the young mathematicians, and a period of study there would be most stimulating and profitable for me, especially if I could manage to get on the good side of Poincaré."

Klein himself in his youth had made the trip to Paris in the company of his friend Sophus Lie, and both he and Lie had brought away their knowledge of group theory, which had played an important part in their careers. Now, according to Hurwitz, Klein always tried to send promising young German mathematicians to Paris.

Hurwitz himself seconded Klein's recommendation: "I fear the young talents of the French are more intensive than ours, so we must master all their results in order to go beyond them."

By the end of March 1886, Hilbert was on his way.

IV

Paris

On the train to Paris, Hilbert had the good luck to be in the same compartment with a student from the École Polytechnique who knew all the French mathematicians "at least from having looked at them." But in Paris, of necessity, he had to join forces with the disagreeable Study, who was already established there, also on Klein's recommendation.

Together, Hilbert and Study paid the mathematical visits which Klein had recommended. When they wrote to Klein, they read their letters aloud to one another so that they would not repeat information.

As soon as Hilbert was settled, he wrote to Klein. The letter shows how important he considered the professor. It was carefully drafted out with great attention to the proper, elegant wording, then copied over in a large, careful Roman script rather than the Gothic which he continued to use in his letters to Hurwitz.

"The fact that I haven't allowed myself at an earlier point in time to entrust the international post with a letter to you is due to the various impediments and the unforeseen cares which are always necessary on the first stay in a strange country. Fortunately, I have now adjusted to the climate and accommodated myself to the new environment well enough that I can start to spend my time in the way that I wish . . . ."

He tried hard to follow Klein's instructions about becoming friendly with Poincaré. The Frenchman was eight years older than he. Already he had published more than a hundred papers and would shortly be proposed for the Academy with the simple statement that this work was "above ordinary praise." In his first letter, Hilbert reported to Klein that Poincaré had not yet returned the visit which he and Study had paid on him; however, he added, he had heard him lecture at the Sorbonne on potential theory and the mechanics of fluids and had later been introduced to him.

"He lectures very clearly and to my way of thinking very understandably although, as a French student here remarks, a little too fast. He gives the impression of being very youthful and a bit nervous. Even after our introduction, he does not seem to be very friendly; but I am inclined to attribute this to his apparent shyness, which we have not yet been in a position to overcome because of our lack of linguistic ability."

By the time that Hilbert wrote to Klein again, Poincaré had returned the visit of the young Germans. "But about Poincaré I can only say the same – that he seems reserved because of shyness, but that with skillful treatment he would open up."

In replying to the letters from Paris, Klein (who was now established in Göttingen) played no favorites between his two young mathematicians. "It is thoroughly necessary that you and Hilbert have personal contact with Gordan and Noether," he wrote to Study. "Next time," he concluded his letter, "I shall write to Dr. Hilbert." But Hilbert seems to have placed a higher value on Klein's letters, for he preserved those written to Study as well as those written to himself.

The French mathematicians – Hilbert wrote to Klein – welcomed him and Study with great warmth. Jordan was most kind "and he is the one who presents the most devoted greetings to you." He gave a dinner for Hilbert and Study "to which only Halphen, Mannheim and Darboux were invited." Since, however, everyone spoke German in deference to the visitors, the conversation on mathematics was "very superficial."

Hilbert was not impressed with the mathematical lectures he heard. "French students do not have much that would interest us." Picard's lectures seemed "least elementary." Although Hilbert found Picard's pronunciation hard to understand, he attended his lectures regularly. "He gives the impression of being very energetic and positive in his conversation as well as in his teaching."

Some of the well-known mathematicians were a disappointment. "Concerning Bonnet – the trouble we went through to find him – a cruel fate sent us to three different houses first – was scarcely in proportion to the advantage we expected from such an old mathematician. He is obviously no longer responsive to mathematical things."

Hilbert and Study attended the meeting of the Société Mathématique in the hope of becoming acquainted with some young – or, at least, younger – mathematicians: "one reason is to be not always towered over by men who are so much greater than we are." Among those they met, Hilbert found Maurice d'Ocagne especially outstanding "because of his pleasant manners and approachability." In the course of the Société meeting he saw how he could sketch out a more direct proof than the one given by d'Ocagne of a theorem in his communication. "So I took courage, supported by Halphen, to point out this way of proving it." D'Ocagne asked Hilbert to write down his proof and offered to correct the French in case he wanted to publish it in the *Comptes Rendus*. "But I do not want to go into this thing because I think neither the theorem itself nor the proof is important enough to be put in the *Comptes Rendus*."

"As to the publications of Poincaré," Klein commented in this connection, "I have always the impression that there is the intention to publish something even if none or few new results are present. Do you approve of this? Have you happened to hear in Paris that people there have the same opinion?"

Among the French mathematicians, it was Hermite who seemed the most attractive to Hilbert:

"He not only showed us all his well-known politeness by returning our visit promptly, but also showed himself very kind . . . by offering to spend the morning with me when he doesn't have a lecture."

The young Germans went back for a second visit. Hermite seemed very old to them — he was 64 — "but extraordinarily friendly and hospitable." He talked of his law of reciprocity in binary forms and encouraged them to extend it to ternary forms. In fact, much of the conversation was on invariants, since that was the subject in which Hermite knew his young visitors were most interested. He directed their attention to the most famous unsolved problem in the theory — what was known as "Gordan's Problem" after Klein's friend Gordan at Erlangen — and told them at length about his correspondence with Sylvester concerning the latter's efforts to solve it.

"The way Hermite talked about other, non-scientific topics proves that he has kept his youthful attitude into old age," Hilbert wrote admiringly to Klein.

While Hilbert was having these stimulating contacts in Paris, Minkowski was still soldiering in Königsberg. "I have stood guard duty at 20°, and been forgotten to be relieved on Christmas Eve . . . ." But he was hoping soon "to renew an old acquaintance with Frau Mathematika," and begged for news in the smallest detail of all that had happened to his friend "in enemy territory."

"And if one of the great gentlemen, Jordan or Hermite, still remembers me, please give him my best regards and make it clear that I less through nature than through circumstance am such a lazybones."

In Paris, Hilbert was concentrating singlemindedly on mathematics. The letters to Klein record no sightseeing expeditions but mention only his desire to visit the observatory. In addition to the mathematical calls and lectures, he was attempting to edit and copy "in pretty writing" the paper for his habilitation. The work was progressing well.

At the end of April 1886, Study went home to Germany and reported to Klein in person on the activities in Paris.

"Not as much about mathematics as I expected," Klein commented disapprovingly to Hilbert. He then proceeded to fire off half a dozen questions and comments which had occurred to him while glancing through the most recent number of the *Comptes Rendus:* "Who is Sparre? The so-called Theorem of M. Sparre is already in a Munich dissertation (1878, I think). Who is Stieltjes? I have an interest in this man. I have come across an earlier paper by Humbert – it would be very interesting if you could check on the originality of his work (perhaps via Halphen?) and find out for me a little more about his personality. It is strange that the geometry in the style of Veronese-Segré happens to be coming back in fashion again..." The tone was more intimate than when Klein had been writing to the two young men together. "Hold it always before your eyes," he admonished Hilbert, "that the opportunity you have now will never come again."

This letter of Klein's found poor Hilbert spending a miserable month. The doctor diagnosed an illness of acclimatization "while I think it is a terrible poisoning of the stomach from H_2SO_4, which one has to drink here in a thin and pallid form under the name of wine." There were no more calls and the copying of the habilitation paper had to be postponed. He managed only to drag himself to lectures and meetings. "Everything stops when the inadequacy of the human organism shows itself . . . ."

He may also have been just a little homesick.

By the end of June, on his way back to Königsberg, he was happy and full of enthusiasm. He stopped in Göttingen and reported to Klein on the Paris experiences. It was his first visit to the University, and he found himself charmed by the little town and the pretty, hilly countryside, so different from the bustling city of Königsberg and the flat meadows beyond it. He also stopped in Berlin, where he "paid a visit to everything that has anything to do with mathematics." This included even the formidable Leopold Kronecker.

Kronecker was a tiny man, scarcely five feet tall, who had so successfully managed his family's business and agricultural affairs that he had been able to retire at the age of 30 and devote the rest of his life to his hobby, which was mathematics. As a member of the Berlin Academy, he had regularly taken advantage of his prerogative to deliver lectures at the University. He was now 63 and only recently, since the retirement of Kummer, had he become an official professor.

Kronecker had made very important contributions, especially to the higher algebra; but he once remarked that he had spent more time thinking about philosophy than about mathematics. He was now disturbing his fellow mathematicians, particularly in Germany, by his loudly voiced doubts about the soundness of the foundations of much of the contemporary mathematics. His principal concern was the concept of the arithmetic continuum, which lies at the foundation of analysis. The continuum is the totality of real numbers – positive and negative – integers, fractions or rationals, and irrationals – which provides mathematicians with a unique number for every point on a line. Although the real numbers had been used for a long time in mathematics, it was only during the current century that their nature had been clarified in a precise and rigorous manner in the work of Cauchy and Bolzano and, more recently, of Cantor and Dedekind.

The new formulation did not satisfy Kronecker. It was his contention that nothing could be said to have mathematical existence unless it could actually be constructed with a finite number of positive integers. In his view, therefore, common fractions exist, since they can be represented as a ratio of two positive integers, but irrational numbers like π do not exist – since they can be represented only by an infinite series of fractions. Once, discussing with Lindemann the proof that π is transcendental, Kronecker objected: "Of what use is your beautiful investigation regarding π? Why study such problems when irrational numbers do not exist?" He had not yet made his remark that "God made the natural numbers, all else is the work of man," but already he was talking confidently of a new program which would "arithmetize" mathematics and eliminate from it all "non-constructive" concepts. "And if I can't do this," he said, "it will be done by those who come after me!"

Although a man of many admirable qualities, Kronecker had been virulent and very personal in his attacks on the men whose mathematics he disapproved. ("In fact," recalled Minkowski in a letter to Hilbert, "I did not hear much good about Kronecker even when I was in Berlin.") The distinguished old Weierstrass had been reduced almost to tears by Kronecker's remarks about "the incorrectness of *all* those conclusions with which *so-called* analysis works at present." The high-strung, sensitive

Cantor, as a result of Kronecker's attacks on the theory of sets, had broken down completely and had had to seek asylum in a mental institution.

Hilbert had been warned not to expect a welcome from Kronecker, but surprisingly he was received – he wrote to Klein – "in a very friendly way."

Back home in Königsberg, he settled down to the serious business of habilitation. The work he had prepared was a much more ambitious paper than the doctoral dissertation had been, but still on the subject of invariant theory. To a later mathematician, who studied "every line" of Hilbert's work during his own student days, the habilitation paper was to seem a curiously false start: "He begins with the claim that it is a most important point of view, then it just goes out like a burnt match. Nothing came of it . . . . I was always surprised that for several years Hilbert went around in a direction that didn't lead anywhere, perhaps because of the too formal point of view which he took; and this may have been partly due to his contact with Study."

In addition to his paper, the candidate for habilitation also had to deliver a lecture on a topic selected by the faculty from a choice which he offered. Hilbert proposed "The Most General Periodic Functions" and "The Concept of the Group." The faculty selected the first topic, which was also the one which he himself preferred. The lecture was presented to the satisfaction of everyone concerned; the colloquium examination, passed successfully. On July 8, 1886, Hilbert was able to write to Klein: "The title with which you undeservedly addressed me in your last letter is now in actuality mine."

Earlier, there had been some discussion between Hilbert and Klein over the advisability of Hilbert's habilitating at Königsberg. The East Prussian capital was very much on the outskirts of mathematical activity. Few mathematics students were willing to come that far, so few, in fact, that Lindemann had to refuse Minkowski's request to habilitate at Königsberg when he got out of the army.

"But, after all, I am content and full of joy to have decided myself for Königsberg," Hilbert wrote to Klein. "The constant association with Professor Lindemann and, above all, with Hurwitz is not less interesting than it is advantageous to myself and stimulating. The bad part about Königsberg being so far away from things I hope I will be able to overcome by making some trips again next year, and perhaps then I will get to meet Herr Gordan . . . ."

Almost half of the great creative years between twenty and thirty were gone.

V

Gordan's Problem

Hilbert was resolved that as a docent he would educate himself as well as his students through his choice of subjects and that he would not repeat lectures, as many docents did. At the same time, on the daily walk to the apple tree, he and Hurwitz set for themselves the goal of "a systematic exploration" of mathematics.

The first semester he prepared lectures on invariant theory, determinants and hydrodynamics, the last at the suggestion of Minkowski, who was habilitating at Bonn and showing an interest in mathematical physics. There were not many who took advantage of this earliest opportunity to hear David Hilbert. Only in the lectures on invariant theory was he able to draw the number of students required by the University for the holding of a class. "Eleven docents depending on about the same number of students," he complained disgustedly to Minkowski. In honor of his new status he had a formal picture taken. It showed a young man with glasses, a somewhat straggly moustache and already thinning hair, who looked as if he might be expected to go after what he wanted.

In Bonn, Minkowski was having his troubles. He did not find the other docents congenial, and the mathematics professor had been taken ill. "I feel his absence especially. He was the only one here to whom I could put a mathematical question, or with whom I could speak at all on a mathematical subject." Whenever he had the opportunity, he returned to Königsberg and joined Hilbert and Hurwitz on their daily walks.

During these years the friendship between Hilbert and Minkowski deepened. Minkowski was a frequent vacation guest at Rauschen. Receiving the photograph of Hilbert after one of the Rauschen visits, Minkowski wrote, "If I had not seen you in it so stately and dignified, I would otherwise have had to think of the outlandish impression which you made on me in your Rauschen outfit and hairstyle at our brief meeting this summer." He added, musingly: "That we, although so close, could not at all open up to one another was for me more than a little surprising."

In their correspondence they continued to address each other by the formal pronoun "Sie"; but Hilbert, sending Minkowski a reprint of his first published work — the paper which Klein had presented to the Leipzig Academy the previous year — inscribed it: "To his friend and colleague in the closest sense . . . from the author."

That first year as a docent, Hilbert made none of the trips which he had so optimistically planned in order to compensate himself for the isolation of Königsberg. Later he was to recall the years in the "security" of his native city as a time of "slow ripening." The second semester he gave the lectures on determinants and hydrodynamics which he had originally hoped to give the first semester. He began to plan lectures on spherical harmonics and numerical equations. In spite of the variety of his lectures, his own published work continued to be entirely in the field of algebraic invariants; but he also interested himself in questions in other fields.

Finally, at the beginning of 1888, he felt that he was at last ready to take the trip which he had so long promised himself. He drew up an itinerary which would allow him to call on 21 prominent mathematicians, and in March he set out. In his letters to Minkowski he jokingly referred to himself as "an expert invariant-theory man." Now he went first to Erlangen, where the "king of the invariants" held his court.

Paul Gordan was an impressive personality among the mathematicians of the day. Twenty-five years older than Hilbert, he had come to science rather late. His merchant father, while recognizing the son's unusual computational ability, had refused for a long time to concede his mathematical ability. A one-sided, impulsive man, Gordan was to leave a curiously negative mark upon the history of mathematics; but he had a sharp wit, a deep capacity for friendship, and a kinship with youth. Walks were a necessity of life to him. When he walked by himself, he did long computations in his head, muttering aloud. In company he talked all the time. He liked to "turn in" frequently. Then, sitting in some cafe in front of a foaming stein of the famous Erlangen beer, surrounded by young people, a cigar always in his hand, he talked on, loudly, with violent gestures, completely oblivious of his surroundings. Almost all of the time he talked about the theory of algebraic invariants.

It had been Gordan's good fortune to enter this theory just as it moved onto a new level. The first years of development had been devoted to determining the laws which govern the structure of invariants; the next concern had been the orderly production and enumeration of the invariants, and this was Gordan's meat. Sometimes a piece of his work would contain nothing but formulas for 20 pages. "Formulas were the indispensable supports for the formation of his thoughts, his conclusions and his mode of expression," a friend later wrote of him. Gordan's strength, however, in the invention and execution of the formal algebraic processes was considerable. At the beginning of his career, he had made the first break-through in a famous invariant problem. For this he had been awarded his title as king of the invariants. The general problem, which was still unsolved and now the most famous problem in the theory, was called in his honor "Gordan's Problem." This was the problem which Hermite had discussed with Hilbert and Study in Paris.

"Gordan's Problem" was far removed from the "solving for *x*" with which algebra had begun so many centuries before. It was a sophisticated "pure mathematical" question posed, not by the physical world, but by mathematics itself. The internal structure of all invariant forms was by this time known. Although there would be certain ambiguities and repetitions, different invariant forms of specified order and degree could be written down and counted, at least in principle. The next question was of a quite different nature, for it concerned the totality of invariants. Was there a *basis*, a finite system of invariants in terms of which all other invariants, although infinite in number, could be expressed rationally and integrally?

Gordan's great achievement, exactly 20 years before the meeting with Hilbert, had been to prove the existence of a finite basis for the binary forms, the simplest of all algebraic forms. Characteristically, his proof had been a computational one, based on the nature of certain elementary operations which generate invariants. Today it is dismissed as "crude computation"; but that it was, in its day, a high point in the history of invariant theory is apparent from the fact that in 20 years of effort by English, German, French and Italian mathematicians, no one had been able to extend Gordan's proof beyond binary forms, although in certain specific cases the theorem was known to be true. The title won in 1868 remained unchallenged. Just before Hilbert's arrival in Erlangen, Gordan had published the second part of his "Lectures on Invariant Theory," the plan of this work being primarily "to expound and exemplify worthily" (as a writer of the day explained) the theorem which he had proved at that time.

Hilbert had been familiar with Gordan's Problem for some time; but now, listening to Gordan himself, he seems to have experienced a phenomenon which he had not experienced before. The problem captured his imagination with a completeness that was almost supernatural.

Here was a problem which had every one of the characteristics of a great fruitful mathematical problem as he himself was later to list them:

Clear and easy to comprehend ("for what is clear and easily comprehended attracts, the complicated repels").

Difficult ("in order to entice us") *yet not completely inaccessible* ("lest it mock our efforts").

Significant ("a guidepost on the tortuous paths to hidden truths").

The problem would not let him go. He left Gordan, but Gordan's Problem accompanied him on the train up to Göttingen, where he went to visit Klein and H. A. Schwarz. Before he left Göttingen, he had produced a shorter, more simple, more direct version of Gordan's famous proof of the theorem for binary forms. It was, according to an American mathematician of the period, "an agreeable surprise to learn that the elaborate proofs of Gordan's theorem formerly current could be replaced by one occupying not more than four quarto pages."

From Göttingen, Hilbert went on to Berlin and visited Lazarus Fuchs, who was now a professor at the university there; also Helmholtz; and Weierstrass, who had recently retired. He then paid another call on Kronecker. He had a great deal of admiration for Kronecker's mathematical work, but still he found the older man's authoritarian attitude toward the nature of mathematical existence extremely distasteful. Now he discussed with Kronecker some plans for future investigations in invariant theory. Kronecker does not seem to have been much impressed. He cited a work of his own and said, Hilbert noted, "that my investigation on the subject is contained therein." They had a long talk, however, about Kronecker's ideas on what constitutes mathematical existence and his objections to Weierstrass's use of irrational numbers. "Equal is only 2 = 2. . . . Only the discreet and singular have significance," Hilbert wrote in the little booklet in which he kept notes on the conversations with the mathematicians he visited. The importance the conversation with Kronecker had in Hilbert's mind at this time is indicated by the fact that he devoted four pages of his notebook to it — the other mathematicians visited, including Gordan, never received more than a page.

He left Kronecker, still thinking about Gordan's Problem.

Back home in Königsberg, the problem was with him in the midst of pleasure and work, even at dances, which he loved to attend. In August he went up to Rauschen, as was still his custom; and from Rauschen, on September 6, 1888, he sent a short note to the *Nachrichten* of the Göttingen Scientific Society. In this note he showed in a totally unexpected and original way how Gordan's Theorem could be established, by a uniform method, for forms in any desired number of variables.

No one was prepared for the announcement of the solution of the famous old problem, and the first reaction was almost sheer disbelief.

Since Gordan's own solution of the simplest case, the solution of the general problem had been sought in essentially the same manner, by means of the same kind of elaborate algorithmic apparatus which had been used so successfully by Gordan. With many variables and a complicated transformation group, this approach became fantastically difficult. It was not unusual for a single formula to run from page to page in the *Annalen*. "Comparable only to the formulas which describe the motion of the moon!" a later mathematician complained. In this atmosphere of absolute formalism it had occurred to Hilbert that the only way to achieve the desired proof would be to approach it from a path entirely different from the formalistic one which all investigators to date had taken and found impenetrable. He had set aside the whole elaborate apparatus and rephrased the question essentially as follows:

"If an infinite system of forms be given, containing a finite number of variables, under what conditions does a finite set of forms exist, in terms of which all the others are expressible as linear combinations with rational integral functions of the same variables for coefficients?"

The answer he came to was that such a set of forms *always* exists.

The foundation on which this sensational proof of the existence of a finite basis of the invariant system rested was a lemma, or auxiliary theorem, about the existence of a finite basis of a module, a mathematical idea he had obtained from the study of Kronecker's work. The lemma was so simple that it seemed almost trivial. Yet the proof of Gordan's general theorem followed directly from it. The work was the first example of the characteristic quality of Hilbert's mind — what one of his pupils was to describe as "a natural naiveté of thought, not coming from authority or past experience."

When the proof of Gordan's Theorem appeared in print in December, Hilbert promptly fired off a copy to Arthur Cayley, who half a century before had laid the foundation of the theory. ("The theory of algebraic invariants," a later mathematician once wrote, "came into existence somewhat like Minerva: a grown-up virgin, mailed in the shining armor of algebra, she sprang forth from Cayley's jovian head. Her Athens, over which she ruled and which she served as a tutelary and beneficent goddess, was projective geometry. From the beginning she was dedicated to the proposition that all projective coordinate systems are created equal ....")

"Dear Sir," Cayley replied politely from Cambridge on January 15, 1889, "I have to thank you very much for the copy of your note .... It [seems] to me that the idea is a most important valuable one, and that it ought to lead to a demonstration of the theorem as to invariants, but I am unable to satisfy myself as yet that you have obtained such a demonstration."

By January 30, however, having received two explanatory letters from Hilbert in the intervening time, Cayley was congratulating the young German: "My difficulty was an *a priori* one, I thought that the like process should be applicable to semi-invariants, which it seems it is not; and now I quite see.... I think you have found the solution of a great problem."

Hilbert had solved Gordan's Problem very much as Alexander had untied the Gordian Knot.

At Gordium [Plutarch tells us] he saw the famous chariot fastened with cords made of the rind of the cornel-tree, which whosoever should untie, the inhabitants had a tradition, that for him was reserved the empire of the world. Most authors tell the story that Alexander, finding himself unable to untie the knot, the ends of which were secretly twisted round and folded up within it, cut it asunder with his sword. But Aristobulus tells us it was easy for him to undo it, by only pulling the pin out of the pole, to which the yoke was tied, and afterwards drawing of the yoke itself from below.

To prove the finiteness of the basis of the invariant system, one did not actually have to construct it, as Gordan and all the others had been trying to do. One did not even have to show how it could be constructed. All one had to do was to prove that a finite basis, of logical necessity, *must exist*, because any other conclusion would result in a contradiction — and this was what Hilbert had done.

The reaction of some mathematicians was similar to what must have been the reaction of the Phrygians to Alexander's "untying" of the knot. They were not at all sure that he had untied it. Hilbert had not produced the basis itself, nor had he given a method of producing it. His proof of Gordan's Theorem could not be utilized to produce in actuality a finite basis of the invariant system of even a single algebraic form.

Lindemann found his young colleague's methods "unheimlich" — *uncomfortable, sinister, weird.* Only Klein seemed to recognize the power of the work — "wholly simple and, therefore, logically compelling" — and it was at this time that he decided he must get Hilbert to Göttingen at the first opportunity. Gordan himself announced in a loud voice that has echoed in mathematics long after his own mathematical work has fallen silent:

"Das ist nicht Mathematik. Das ist Theologie."

Hilbert had now publicly taken a position in the current controversy provoked by Kronecker over the nature of mathematical existence. Kronecker insisted that there could be no existence without construction. For him, as for Gordan, Hilbert's proof of the finiteness of the basis of the invariant system was simply not mathematics. Hilbert, on the other hand, throughout his life was to insist that if one can prove that the attributes assigned to a concept will never lead to a contradiction, the mathematical existence of the concept is thereby established.

In spite of the philosophical difference, Hilbert was at this time greatly under the influence of the mathematical ideas of Kronecker — in fact, the fundamental significance of his work in invariants was later to be seen as the application of arithmetical methods to algebraic problems. He sent a copy of every paper he published to Kronecker. Nevertheless, Kronecker remarked petulantly to Minkowski that he was going to stop sending papers to Hilbert if Hilbert did not send papers to him. Hilbert promptly composed a letter which managed to be formal and respectful but firm:

"I remember exactly, and my list of mailed papers also shows it clearly, that I have taken the liberty of sending you a copy of each paper without exception immediately after its publication; and you have had the kindness to send your thanks on postcards for some of the last mailings. On the other hand, most honorable professor, it has never happened that a reprint of one of your papers has arrived as a gift from you to me. When I had the honor of calling on you about a year ago, however, you mentioned that you would choose something from your papers and send it to me. Under the circumstances I believe that there must be some misunderstanding, and I write these lines to remove it as fast and as surely as possible."

Then, with many crossings-out, he struggled to express the idea that what he had written should not be construed as expressing any other meaning than the stated one: not reproaches, but just explanations. He finally gave up, and simply signed himself, "Most respectfully, David Hilbert."

During the next two years, Hilbert, still a docent, sent two more notes to the *Nachrichten* and then in 1890 brought all his papers on algebraic forms together into a unified whole for the *Annalen*. By this time the revolutionary effect of Hilbert's work was being generally recognized and accepted. Gordan, offering another proof of one of Hilbert's theorems, was deferential to the young man — Herr Hilbert's proof was "completely correct," he wrote, and his own proof would not even have been possible "if Herr Hilbert had not utilized in invariant theory concepts which had been developed by Dedekind, Kronecker and Weber in another part of mathematics."

While Hilbert was thus involved in the purest of pure mathematics, Minkowski was moving increasingly away from it. Heinrich Hertz, two years after his discovery of the electromagnetic waves predicted by Maxwell, and still only 31 years old, had recently become professor of physics at Bonn. Minkowski, complaining of "a complete lack of half-way normal mathematicians" among his colleagues, found himself attracted more and more by Hertz and by physics. At Christmas he wrote that, contrary to his custom, he would not be spending the vacation at Königsberg:

"I do not know if I need console you though, since this time you would have found me thoroughly infected with physics. Perhaps I even would have had to pass through a 10-day quarantine period before you and Hurwitz would have admitted me again, mathematically pure and unapplied, to your joint walks."

At another time he wrote:

"The reason that I am now almost completely swimming in physical waters is because here at the moment as a pure mathematician I am the only feeling heart among wraiths. So for now," he explained, "in order to have points in common with other mortals, I have surrendered myself to magic — that is to say, physics. I have my laboratory periods at the Physics Institute; at home I study Thomson, Helmholtz and consorts. And from the end of next week on, I will even work several days a week in a blue smock in an institute for the production of physical instruments, a technician, therefore, and as practical as you can imagine!"

But the diverging of scientific interests did not affect the friendship; and, in fact, it was at this time that the two young men made the significant transition in their correspondence from the formal pronoun "Sie" to the intimate "du."

The Privatdozent years seemed to stretch out interminably. The letters were much concerned with the possibility of promotion. In 1891 Minkowski wrote that he had been told that he might be proposed for a position in Darmstadt. "But this ray of hope could easily shine so long that it shines upon mostly grey hair." That same year — apparently with special permission from the University — Hilbert was delivering his lectures on analytic functions to only one student — an American from Baltimore — a man somewhat older than the young lecturer but, in his opinion, "very sharp and extraordinarily interested." This was Fabian Franklin, an important man in invariant theory and the successor of Sylvester at Johns Hopkins.

Because there were few mathematics students at Königsberg, Hilbert attended the meetings of the natural scientists as well as those of the mathematicians. But Königsberg was surprisingly full of congenial young people. Wiechert was a docent too; and he had recently been joined by a student named Arnold Sommerfeld, with whom he was devising a harmonic analyzer. Both Wiechert and Sommerfeld were to become masters of electrodynamic theory, but when "Little Sommerfeld" heard Hilbert lecture on ideal theory, he became convinced that his interest lay entirely with the most pure and abstract mathematics. "Already," he later commented, "it was clear that a spirit of a special sort was at work."

There was lots of happy social life. Hilbert was a gay young man with a reputation as a "snappy dancer" and a "charmeur," according to a relative. He flirted, outrageously, with a great number of girls. His favorite partner for all activities, however, was Käthe Jerosch, the daughter of a Königsberg merchant, an outspoken young lady with an independence of mind that almost matched his own.

Even after the work of 1890, Gordan's Problem still would not let Hilbert go. As a mathematician he preferred an actual construction to a proof of existence. "There is," as one mathematician has said, "an essential difference between proving the existence of an object of a certain type by constructing a tangible example of such an object, and showing that if none existed one could deduce contrary results. In the first case one has a tangible object, while in the second one has only the contradiction." He would very much have liked to produce for old Kronecker, Gordan and the rest a constructive proof of the finiteness of the basis of the invariant system. At the moment there was simply no method at hand.

In the course of the next two years, however, the nature of his work began to change. It became infused with the ideas of algebraic number fields. Again, Kronecker's ideas were important. And it was here that Hilbert found at last the powerful new tools he had been seeking. In a key work, in 1892, he took up the question of exactly what was needed to produce in actuality a full system of invariants in terms of which all the other invariants could be represented. Using as a foundation the theorem which he had earlier proved, he was able to produce what was in essence a finite means of executing the long sought construction.

Although Hilbert was not the first to make use of indirect, non-constructive proofs, he was the first to recognize their deep significance and value and to utilize them in dramatic and extremely beautiful ways. Kronecker had recently died; but to those who like Kronecker still declared that existence statements are meaningless unless they actually specify the object asserted to exist, Hilbert was always to reply:

"The value of pure existence proofs consists precisely in that the individual construction is eliminated by them, and that many different constructions are subsumed under one fundamental idea so that only what is essential to the proof stands out clearly; brevity and economy of thought are the *raison d'être* of existence proofs.... To prohibit existence statements ... is tantamount to relinquishing the science of mathematics altogether."

Now, through a proof of existence, Hilbert had been able to obtain a construction. The impetus which his achievement gave to the use of existential methods can hardly be overestimated.

Minkowski was utterly delighted:

"For a long while it has been clear to me that it could be only a question of time until the old invariant question was settled by you — only the dot was lacking on the 'i'; but that it all turned out to be so surprisingly simple has made me very happy, and I congratulate you."

He was inspired to literary flight and an assortment of metaphors. The first existence proof might have got smoke in Gordan's eyes, but now Hilbert had found a smokeless gunpowder. The castle of the robber barons — Gordan and the rest — had been razed to the ground with the danger that it might never rise again. Hilbert would be doing a service to his fellow mathematicians if he would bring together the materials in this area on which one could rebuild. But he probably would not want to spend his time doing that. There were still too many other things that he was capable of doing!

Gordan himself conceded gracefully.

"I have convinced myself that theology also has its merits."

When Klein went to Chicago for what was billed as an "International Congress of Mathematicians" to celebrate the founding of the University of Chicago, he took with him a paper by Hilbert in which that young man matter-of-factly summarized the history of invariant theory and his own part in it:

"In the history of a mathematical theory the developmental stages are easily distinguished: the naive, the formal, and the critical. As for the theory of algebraic invariants, the first founders of it, Cayley and Sylvester, are together to be regarded as the representatives of the naive period: in the drawing up of the simplest invariant concepts and in the elegant applications to the solution of equations of the first degrees, they experienced the immediate joy of first discovery. The inventors and perfecters of the symbolic calculation, Clebsch and Gordan, are the champions of the second period. The critical period finds its expressions in the theorems I have listed above . . . ."

The theorems he referred to were his own.

It was a rather brash statement for a young mathematician who was still not even an Extraordinarius, but it had considerable truth in it. Cayley and Sylvester were both alive, one at Cambridge and the other at Oxford. Clebsch was dead, but Gordan was one of the most prominent mathematicians of the day. Now suddenly, in 1892, as a result of Hilbert's work, invariant theory, as it had been treated since the time of Cayley, was finished. "From the whole theory," a later mathematician wrote, "the breath went out."

With the solution of Gordan's Problem, Hilbert had found himself and his method — an attack on a great individual problem, the solution of which would turn out to extend in significance far beyond the problem itself. Now something totally unexpected occurred. The problem which had originally aroused his interest had been solved. The solution released him.

At the conclusion of his latest paper on invariants he had written: "Thus I believe the most important goals of the theory of function fields generated by invariants have been obtained." In a letter to Minkowski, he announced with even more finality: "I shall definitely quit the field of invariants."

VI

Changes

During the next three years Hilbert rose in the academic ranks, did all the things that most young men do at this time of their lives, married, fathered a child, received an important assignment, and made a decision which changed the course of his life.

This sudden series of events was set into motion by the death of Kronecker and the game of "mathematical chairs" which ensued in the German universities. Suddenly it seemed that the meager docent years might be coming to an end. Minkowski calling in Berlin on Friedrich Althoff, who was in charge of all matters pertaining to the universities, heralded the news:

"A. says . . . the following are supposed to receive paid Extraordinariats: you, I, Eberhard, and Study. I have not neglected to represent you to A. as the coming man in mathematics. . . . As to Study, in conscience I could only praise his good intentions and his diligence. A. is very devoted to you and Eberhard."

At almost the same time Hurwitz, who had been an associate professor (Extraordinarius) at Königsberg for eight years, received an offer of a full professorship from the Swiss Federal Institute of Technology in Zürich. This meant an end to the daily mathematical walks, but opened up the prospect of Hilbert's being appointed to Hurwitz's place.

"Through this circumstance," Minkowski wrote affectionately, "your frightful pessimism will have been allayed so that one dares again to venture a friendly word to you. In some weeks now, hopefully, the Privatdozent-sickness will be definitely over. You see – at last comes a spring and a summer."

In June, Hurwitz married Ida Samuels, the daughter of the professor of medicine. Hilbert had recently become engaged to Käthe Jerosch, and after Hurwitz's wedding he was increasingly impatient with the slow pace of promotion. At last, in August, the faculty unanimously voted him to succeed to Hurwitz's place. He announced the setting of the date of his wedding at the same time he communicated the news of the promotion to Minkowski.

Minkowski replied happily with his congratulations: "You will now have finally been converted to the idea that those in the decisive positions are sincerely well disposed toward you. Your prospects for the future, therefore, are excellent."

The Hilbert and Jerosch families had long been friends. From the outset it was generally agreed that Hilbert had found the perfect mate for himself. "She was a full human being in her own right, strong and clear," one of Hilbert's earliest pupils wrote of Käthe, "and always stood on the same footing with her husband, kindly and forthright, always original."

A photograph, taken about this time, shows the young couple. He is 30; she is 28. Already they look rather like one another. They are almost the same height, mouths wide and firm, strong noses, a level clear-eyed look. Hilbert's head seems relatively small. He has grown a beard. Already his hair has receded until the high scholar's forehead stands out impressively. Neither pretty nor homely, Käthe has good features, but she seems more interested in things other than her own appearence. Her dark hair is parted in the middle, drawn back rather severely, and coiled on the top of her head toward the back.

On October 12, 1892, Hilbert and Käthe Jerosch were married.

("The pleasant frame of mind in which you find yourself cannot help but have repercussions in your scientific work," Minkowski wrote. "I expect another great discovery.")

At almost the same time that Hilbert succeeded Hurwitz in Königsberg, Minkowski received his promised associate professorship in Bonn. He had hoped to go somewhere else, but "it will be better for you to remain in Bonn," Althoff told him. By now Heinrich Hertz had been struck down with the illness which was soon to take his life at the age of 37; Minkowski's interest in physics had abated; and he had returned to his first love, the theory of numbers. But later he once said to Hilbert that if "Papa" Hertz had lived he might have become a physicist instead of a mathematician.

Minkowski's approach to number theory was geometrical, it being his aim to express algebraic conjectures about the rational numbers in terms of geometric figures, an approach which frequently made the proofs more obvious. He was deeply absorbed in a book on this new subject, and his letters to Hilbert were filled with his concern about the presentation of his material. All must be "klipp und klar" before it went to the publisher. Although he called Poincaré "the greatest mathematician in the world," he told Hilbert, "I could not bring myself to publish things in the form in which Poincaré publishes them."

The book frequently kept Minkowski from Königsberg at vacation time. Hilbert complained about a lack of mathematical conversation now that Hurwitz was gone. "I am in a much more unhappy situation than you," Minkowski reminded him. "Just as closed off as Königsberg is from the rest of the world, just so closed off is Bonn from all other mathematicians. One is here a pure mathematics Eskimo!"

By the beginning of the new year (1893) Minkowski was happier. The book was half finished, accompanied by praise from Hermite which Hilbert found very touching.

"You are so kind as to call my old research works a point of departure for your magnificent contribution," the old Frenchman wrote to Minkowski, "but you have left them so far behind that they cannot claim now any other merit than to have suggested to you the direction in which you have chosen to proceed."

Hilbert began the year with a new proof of the transcendence of e (first proved by Hermite) and of π (proved by Lindemann). His proof was a considerable improvement over these earlier ones, astonishingly simple and direct. Here was the great work which Minkowski had been anticipating since the previous fall. Receiving it, he sat down and wrote immediately.

"An hour ago I received your note on e and π . . . and I cannot do other than to express to you right away my sincere heartfelt astonishment. . . . I can picture the exhilaration of Hermite upon reading your paper and, as I know the old gentleman, it won't surprise me if he should shortly inform you of his joy that he is still permitted to experience this."

Along with the professional and personal changes in his life, Hilbert was beginning to show a new mathematical interest. "I shall devote myself to number theory from now on," he had told Minkowski after the completion of the last work on invariants. Now he turned to this new subject.

Gauss, as is well known, placed the theory of numbers at the pinnacle of science. He described it as "an inexhaustible storehouse of interesting truths." Hilbert saw it as "a building of rare beauty and harmony." He was as charmed as Gauss had been by "the simplicity of its fundamental laws, the economy of its concepts, and the purity of its truth"; and both men were equally fascinated by the contrast between the obviousness of the many numerical relationships involved and the "monstrous" difficulty of demonstrating them. Yet, in spite of the similarity of their comments, they were talking about two different versions of number theory.

Gauss was praising the classical theory of numbers, which goes back to the Greeks and deals with the relationships which exist among the ordinary whole, or natural, numbers. Most important are those between the prime numbers, called the "building blocks" of the number system, and the other numbers which, unlike the primes, can be divided by some number other than themselves and 1. By Gauss's time the concept of number had been extended far beyond the natural numbers. But Gauss himself had become the first mathematician to extend the notions of number theory itself beyond the rational "field" in which every sum, difference, product and (unlike among the natural numbers) quotient of two numbers is another number in the field. He did this for those numbers of the form $a + b\sqrt{-1}$ where a and b are rational numbers. These numbers also form a field, an algebraic number field, as do the numbers of the form $a + b\sqrt{2}$, and so on; and they are among the fields which are the subject of what is called algebraic number theory. It was this development, the number theory creation of Gauss, which Hilbert praised.

The greatest obstacle to the extension of number theory to algebraic number fields had been the fact that in most algebraic number fields the fundamental theorem of arithmetic, which states that the representation of any number as the product of primes is unique, does not hold. This obstacle had been eventually overcome by Kummer with the invention of "ideal numbers." Since Kummer, two mathematicians with very different mathematical approaches had been at work in algebraic number fields. Even before Hurwitz had left for Zürich, he and Hilbert had been devoting their daily walks to discussions of the modern number theory works of these two. "One of us took the Kronecker demonstration for the complete factorization in prime ideals and the other took Dedekind's," Hilbert later recalled, "and we found them both abominable." Now he began his work in algebraic number fields in much the same way that he had opened his attack on Gordan's Problem. He went back and thought through the basic idea. His first paper in the new subject was another proof for the unique decomposition of the integers of a field into prime ideals.

Hilbert had scarcely settled down into his new position as an assistant professor with a salary and a wife when there was welcome news. Lindemann had received an offer from Munich and would be leaving Königsberg.

"I take it for granted — and with any sense of justice Lindemann cannot think otherwise — that you should be his successor," Minkowski wrote to Hilbert. "If he succeeds in putting it through, he will at least leave with honor the place which he has occupied for 10 years."

Hilbert of course agreed. The final decision in the matter was not Lindemann's, however, but Althoff's. The faculty nominated Hilbert and three other more established mathematicians for the vacant professorship and sent the list to Berlin.

Althoff was no bureaucrat, but an administrator who had been academically trained. His great goal was to build up mathematics in Germany. He was a good friend of Klein's — the two had served in the army together during the Franco-Prussian War — and he thought very highly of Klein's opinion. Now, from the faculty's impressive list of names, he selected that of the 31-year-old Hilbert. He then proceeded to consult him about the appointment of a successor to his post as Extraordinarius — something almost unheard of.

Here was an opportunity to bring Minkowski back to Königsberg. In spite of the difficult situation which existed at Bonn because of the long illness of the professor of mathematics, Hilbert embarked enthusiastically upon the unfamiliar course of academic diplomacy. He wrote to Minkowski of the possibility that they might soon be together again.

"I would consider it special luck to step into your place at Königsberg," Minkowski replied. "The association with my mathematical colleagues here is really deplorable. One complains of migraine; as for the other, his wife trots in every five minutes in order to give another, non-mathematical direction to the conversation. If I could exchange this association with yours, it would be the difference between day and night for my scientific development."

But the sick professor at Bonn wanted to hang on to Minkowski, to whom he had become accustomed. Althoff liked to keep his professors happy. The negotiations dragged on.

In the meantime, in the new household, things were going along well and according to form. On August 11, 1893, at the seaside resort of Cranz, a first child, a son, was born to the Hilberts and named Franz.

A few weeks after Franz's birth, Hilbert went south to Munich for the annual meeting of the German Mathematical Society, which had recently been organized by a group of mathematicians — Hilbert among them — for the purpose of providing more contact among the different branches of mathematics. At the meeting Hilbert presented two new proofs of the decomposition of the numbers of a field into prime ideals. Although he had only begun to publish in the area of algebraic number theory, his competence apparently impressed the other members. One of the Society's projects was the yearly publication of comprehensive surveys of different fields of mathematics (the first had been on the theory of invariants); and now it was voted that Hilbert and Minkowski, who was of course already well known as a number theorist, be asked to prepare such a report on the current state of affairs in the theory of numbers "in two years." The note of urgency in the assignment was occasioned by the fact that the revolutionary work of Kummer, Kronecker and Dedekind was so extremely complicated or so far in advance of its time that it was still incomprehensible to most mathematicians. That Hilbert and Minkowski could be expected to rectify this situation was a tribute, not only to their mathematical ability, but also to the simplicity and clarity of their mathematical presentation.

That fall the letters that went between Königsberg and Bonn were devoted almost equally to three topics: the organization of the report for the Mathematical Society, the progress of the negotiations to bring Minkowski to Königsberg again, and the fact that baby Franz could already "outshriek" all other babies in his father's opinion.

The situation in Bonn did not improve; by New Year's Day 1894, Minkowski wrote that he had given up almost all hope of obtaining the appointment in Königsberg. Then three days later, following an interview with Althoff, he sent a joyful letter to Hilbert.

"End good, all good . . . . Hearty thanks for all your kind efforts which have led to this happy result; and may we have a pleasant and profitable collaboration which will make the prime numbers and the reciprocity laws *wiggeln und waggeln.*"

On his way up to Königsberg in March, Minkowski stopped in Göttingen. H. A. Schwarz had by this time moved on to Berlin – his place being taken by Heinrich Weber – and Klein had a free hand to put into practice his ideas. Minkowski seems to have been tremendously impressed by the stimulating situation which Klein had already created at the University. "Who knows when I shall have another opportunity to inspire the mathematical workshop which is now of the highest repute?"

With Minkowski's arrival in the spring of 1894, the daily walks to the apple tree and the number theory discussions were happily resumed. It was Hilbert's feeling that he could not have had a better collaborator on the *Zahlbericht*, as the number theory report was called. In spite of Minkowski's mild disposition, he was fundamentally critical, insisted on literary as well as intellectural clarity, "and even to the work of others applied a strict standard."

The *Zahlbericht* now began to take shape in Hilbert's mind. Such an assignment as the one made by the Mathematical Society might be expected to be an unwelcome chore to a young mathematician, but this was not to be the case with Hilbert. Already his own work showed that his particular interest was the extension of the reciprocity laws to algebraic number fields. Now he willingly set aside these plans, seeing in the assigned report an opportunity to lay the foundation needed for deeper investigations. Although he still had no fondness for learning from books, he read everything that had been published on number theory since the time of Gauss. The proofs of all known theorems would have to be weighed carefully. Then he would have to decide in favor of those "the principles of which are capable of generalization and most useful for further research." But before such a selection could be made, the "further research" itself would have to be carried out. The difficulties of thought and style which had barred the way to general appreciation and understanding of his predecessors' work would have to be eliminated. It had been decided that the report should be divided into two parts. Minkowski would treat rational number theory; Hilbert, algebraic number theory. During the year 1894 Hilbert laid the foundations of his share of the *Zahlbericht*.

But, again, the two friends were not to be together for long. Early in December a letter labeled "Very Confidential" arrived from Göttingen.

"Probably you do not yet know," Klein wrote to Hilbert, "that Weber is going to Strassburg. This very evening we will have a meeting of the faculty to choose a committee to set up a list; and as little as I can predict the results, I want to inform you that I shall make every effort to see that no one other than you is called here.

"You are the man whom I need as my scientific complement because of the direction of your work and the power of your mathematical thinking and the fact that you are still in the middle of your productive years. I am counting on it that you will give a new inner strength to the mathematical school here, which has grown continuously and, as it seems, will grow even more — and that perhaps you will even exercise a rejuvenating effect upon me . . . .

"I can't know whether I will prevail in the faculty. I know even less whether the offer will follow from Berlin as we propose it. But this one thing you must promise me, even today: that you will not decline the call if you receive it!"

There is no record that Hilbert ever considered declining. In fact, he wrote to Klein, "Without any doubt I would accept a call to Göttingen with great joy and without hesitation." But he may have had some doubts. Klein was the acknowledged leader of mathematics in Germany. He was a regal man, the word "kingly" being now used most frequently to describe him. Sometimes even "kingly" wasn't strong enough, and one former student referred to him as "the divine Felix." A man who knew him well and was proud of the fact that Klein once took his advice in a personal matter, later confessed that he felt even to the end a distance between himself and Klein "as between a mortal and a god."

As for Klein's feelings – already it was clear that Hilbert questioned any authority, personal or mathematical, and went his own way. Klein was not unaware of the reasons against his choice. When in the faculty meeting his colleagues accused him of wanting merely a comfortable younger man, he replied, "I have asked the most difficult person of all."

Hilbert worked very hard on his reply to Klein's letter, crossing out and rewriting extensively to get exactly the effect he wanted. When he was satisfied, he had Käthe copy his letter in her best handwriting. It was a custom he was to follow often throughout his career.

"Your letter has surprised me in the happiest way," he began. "It has opened up a possibility for the realization of which I might have hoped at best in the distant future and as the final goal of all my efforts....

"Decisive for me above all would be the scientific stimulation which would come from you and the greater sphere of influence and the glory of your university. Besides, it would be the fulfillment of mine and my wife's dearest wish to live in a smaller university town, particularly one which is so beautifully situated as Göttingen."

Upon receiving this letter from Hilbert, Klein proceeded to plan out a campaign.

"I have already told Hurwitz that we will not propose him this time so that we will be more successful in proposing you. We will call Minkowski in second place. I have discussed this with Althoff and he thinks that will make it easier then for Minkowski to get your place in Königsberg."

Within a week he was writing triumphantly to Hilbert:

"This has been just marvellous, much faster than I ever dared to hope it could be. Please accept my heartiest welcome!"

VII

Only Number Fields

The red-tiled roofs of Göttingen are ringed by gentle hills which are broken here and there by the rugged silhouette of an ancient watch tower. Much of the old wall still surrounds the inner town, and on Sunday afternoons the townspeople "walk the wall" – it is an hour's walk. Outside the wall lie the yellow-brick buildings of the Georg August Universität, founded by the Elector of Hannover who was also George II of England. Inside, handsome half-timbered houses line crooked, narrow streets. Two thoroughfares, Prinzenstrasse and Weender Strasse, intersect at a point which the mathematicians call the origin of the coordinates in Göttingen. The center of the town, however, is the Rathaus, or town hall. On the wall of its Ratskeller there is a motto which states unequivocally: *Away from Göttingen there is no life.*

The great scientific tradition of Göttingen derives from Carl Friedrich Gauss, the son of a man who was at different times a gardener, a canal-tender and a brick-layer. Gauss enrolled at the University in the autumn of 1795 as the protégé of the Duke of Brunswick. During the next three years he had so many great mathematical ideas that he could often do no more than record them in his journal. Before he left the University, at the age of 21, he had virtually completed one of the masterpieces of number theory and of mathematics, the *Disquisitiones Arithmeticae.* Later he returned to Göttingen as director of the observatory with incidental duties of instruction. He spent the rest of his life there, leaving his mark on every part of pure and applied mathematics. But when he was an old man and had won a place with Archimedes and Newton in the pantheon of his science, he always spoke of the first years he had spent at Göttingen as "the fortunate years."

Hilbert arrived in Göttingen in March 1895, almost exactly one hundred years after Gauss. It was not immediately apparent to the students that another great mathematician had joined the tradition. Hilbert was too different from the bent, dignified Heinrich Weber whom he replaced and the tall, commanding Klein. "I still remember vividly," wrote Otto Blumenthal, then a student in his second semester, "the strange impression I received of the medium-sized, quick, unpretentiously dressed man with a reddish beard, who did not look at all like a professor."

Klein's reputation drew students to Göttingen from all over the world, but particularly from the United States. The *Bulletin* of the newly founded American Mathematical Society regularly listed the courses of lectures to be given in Göttingen, and at one time the Americans at the University were sufficient in number and wealth to have their own letterhead: *The American Colony of Göttingen.* "There are about a dozen . . . in our lectures," a young Englishwoman named Grace Chisholm (later Mrs. W. H. Young) wrote to her former classmates at Cambridge. "We are a motley crew: five are Americans, one a Swiss-French, one a Hungarian, and one an Italian. This leaves a very small residuum of German blood."

The center of mathematical life was the third floor of the Auditorienhaus. Here Klein had established a reading room, the Lesezimmer, which was entirely different from any other mathematical library in existence at that time. Books were on open shelves and the students could go directly to them. Klein had also established on the third floor what was to become almost his signature: a tremendous collection of mathematical models housed in a corridor where the students gathered before lectures. Although not in actuality a room, it was always referred to as the Room of the Mathematical Models.

Klein's lectures were deservedly recognized as classics. It was his custom often to arrive as much as an hour before the students in order to check the encyclopedic list of references which he had had his assistant prepare. At the same time he smoothed out any roughness of expression or thought which might still remain in his manuscript. Before he began his lecture, he had mapped out in his mind an arrangement of formulas, diagrams and citations. Nothing put on the blackboard during the lecture ever had to be erased. At the conclusion the board contained a perfect summary of the presentation, every square inch being appropriately filled and logically ordered.

It was Klein's theory that students should work out proofs for themselves. He gave them only a general sketch of the method. The result was that a student had to spend at least four hours outside class for every hour spent in class if he wished to master the material. Klein's forte was the comprehensive view. "He possessed the ability to see the unifying idea in far apart problems and knew the art of explaining this insight by amassing the necessary details," a student has said. In the selection of his lecture subjects, Klein pursued a characteristically noble plan: "to gain in the course of time a complete view of the whole field of modern mathematics."

In contrast, Hilbert delivered his lectures slowly and "without frills," according to Blumenthal, and with many repetitions "to make sure that everyone understood him." It was his custom to review the material which he had covered in the previous lecture, a gymnasium-like technique disdained by the other professors. Yet his lectures, so different from Klein's, were shortly to seem to many of the students more impressive because they were so full of "the most beautiful insights."

In a well-prepared lecture by Hilbert the sentences followed one another "simply, naturally, logically." But it was his custom to prepare a lecture in general, and often he was tripped up by details. Sometimes, without especially mentioning the fact, he would develop one of his own ideas spontaneously in front of the class. Then his lectures would be even farther from the perfection of Klein's and exhibit the rough edges, the false starts, the sometimes misdirected intensity of discovery itself.

In the eight and a half years of teaching at Königsberg, Hilbert had not repeated a single subject "with the one small exception" of a one-hour course on determinants. In Göttingen now he was easily able to choose his subjects to adjust to Klein's wishes. The first semester he lectured on determinants and elliptic functions and conducted a seminar with Klein every Wednesday morning on real functions.

Although Hilbert had accepted the professorship in Göttingen with alacrity, there were two aspects of the new situation that bothered him. Käthe was not happy. The society in Göttingen, while more scientifically stimulating for him, lacked the warmth to which she had been accustomed in Königsberg. Carefully observed distinctions of rank cut the professors off from the docents and advanced students. In spite of Klein's kindness, he maintained with the Hilberts, as he did with everyone else, a certain distance. Mrs. Klein (granddaughter of the philosopher Hegel) was a very quiet woman, not the kind who likes to gather people around her. The Klein house at 3 Wilhelm Weber Strasse, big, square and impressive with a bust of Jupiter on the stairs that led to Klein's study, looked already like the institute it was eventually to become. For Hilbert "comradeship" and "human solidarity" were essential to scientific production. Like Käthe, he found the atmosphere at Göttingen distinctly cool.

Hilbert was also concerned, in the beginning, that he might not prove worthy of the confidence which Klein had shown in him. He recognized that he had been taken on faith. Before he had left Königsberg, he had written to Klein, "My positive achievements — which I indeed know best myself — are still very modest." In the draft of a later letter he had returned to this same subject, adding hopefully, "As to my scientific program, I think that I will eventually succeed in shaping the theory of ideals into a general and usable tool (applicable also to analytic functions and differential equations) which will complement the great and promising concept of the group." Then he had carefully crossed out this sentence and noted in the margin: *I have not written this.*

Now, in Göttingen, Hilbert concentrated all his powers on his share of the number theory report for the German Mathematical Society, which he saw as the necessary foundation for his future hopes.

In Königsberg, Minkowski almost immediately received the appointment as his friend's successor. "The whole thing has taken place so quickly that I still have not come to complete consciousness of my astounding luck. In any case, I know I have you alone to thank for everything. I shall see I break out of my cocoon so that no one will hold it against you for proposing me." Minkowski was happy in his new position — professors now went out of their way to describe to him the virtues of their daughters — but since Hilbert's departure, he wrote, he had walked "not once" to the apple tree.

With encouragement from Hilbert, Minkowski now took advantage of the fact that he was a full professor to deliver a course of lectures on Cantor's theory of the infinite. It was at a time when, according to Hilbert, the work of Cantor was still actually "taboo" in German mathematical circles, partly because of the strangeness of his ideas and partly because of the earlier attacks by Kronecker. Although Minkowski admired Kronecker's mathematical work, he deplored as much as Hilbert the way in which the older man had tried to impose his restrictive personal prejudices upon mathematics as a whole.

"Later histories will call Cantor one of the deepest mathematicians of this time," Minkowski said. "It is most regrettable that opposition based not alone on technical grounds and coming from one of the most highly regarded mathematicians could cast a gloom over his joy in his scientific work."

As the year 1895 progressed, the letters between Göttingen and Königsberg became less frequent.

"We both try in silence to crack the difficult and not really very tasty nut of our common report," Minkowski wrote, taking up the correspondence again, "you perhaps with sharper teeth and more exertion of energy."

The idea of the joint report did not really appeal to Minkowski. "I started somewhat too late with my share," he wrote unhappily. "Now I find many little problems it would have been nice to dispose of." He was more interested in his book on the geometry of numbers. "The complete presentation of my investigations on continued fractions has reached almost a hundred printed pages but the all-satisfying conclusion is still missing: the vaguely conceived characteristic criterion for cubic irrational numbers .... But I haven't been able to work on this problem because I have really been working on our report."

Hilbert, on the other hand, was devoting himself wholeheartedly to the report. He was fascinated by the deep connections which had recently been revealed between the theory of numbers and other branches of mathematics. Number theory seemed to him to have taken over the leading role in algebra and function theory. The fact that this had not occurred earlier and more extensively was, in his opinion, due to the disconnected way in which number theory had developed and the fact that its treatment had always been chronological rather than conceptual. Now, he believed, a certain and continuous development could be effected by the systematic building up of the theory of algebraic number fields.

After the Wednesday morning seminars he walked with the students up to a popular restaurant on the Hainberg for lunch and more mathematics. On these excursions he talked freely to them "as equals," according to Blumenthal, but always the subject of conversation at this time was "only algebraic number fields."

By the beginning of 1896, Hilbert's share of the *Zahlbericht* was almost finished, but Minkowski's was not. In February Hilbert proposed that either Minkowski's share should be published with his as it stood, or else it should be published separately the following year.

"I accept your second plan," Minkowski wrote gratefully. "The decision ... is hard on me only insofar as I'll have the guilty feeling for a whole year that I didn't meet the expectations of the Society and, in some degree, your expectations. You, it is true, haven't made any remark of this kind, but .... The reproaches may lose some of their force if now the biggest part of my book is appearing and the rest is following soon. Finally, I can imagine that I am doing what I think is in the interest of the project. I beg you not to think I left you in the lurch."

Within a month after receiving this letter, Hilbert had completed his report on algebraic number fields. It was exactly a year since his arrival in Göttingen. The manuscript, which was to run to almost 400 pages in print, was carefully copied out by Käthe Hilbert in her clear round hand and sent to the printer. The proof-sheets were mailed to Minkowski in Königsberg as they arrived. Minkowski's letters during this period show the affectionate and yet sharp and unrelenting care with which he read them.

"One more remark seems to be necessary on page 204." "I have read till where the long calculations start. They still seem pretty tangled." "This thought is not so simple that it can be silently omitted."

Minkowski had recently received an offer of a position in Zürich. Such an offer, known as "a call," was customarily the subject of complicated ritual and negotiation, since it was the only means by which a man who had become a full professor could further improve his situation. Minkowski had no gift for such parrying. Althoff, he wrote to Hilbert, did not seem eager to keep him at Königsberg. Rather regretfully, he finally accepted the position in Zürich for the fall of 1896.

In Zürich, however, he was again in the company of Hurwitz ("just the same except for a few white hairs"), and the two friends read the remaining proof-sheets of Hilbert's report together. Corrections and suggestions kept coming to Göttingen.

Hilbert began to grown impatient.

Minkowski soothed him: "I understand that you want to be through with the report as soon as possible . . . but as long as there are so many remarks to be made, I can't promise you any great speed . . . ." "A certain care is advisable . . . ." "Comfort yourself with the thought that the report will be finished soon and will gain high approval."

The careful proofreading continued.

By this time Hilbert was beginning to feel more at home in Göttingen. He had found a congenial colleague in Walther Nernst, a professor of physics and chemistry who, like himself, was the son of a Prussian judge. But Hilbert also liked to be with younger people, and now he cheerfully ignored convention in choosing his friends. These included Sommerfeld, who had come to Göttingen to continue his studies and had become Klein's first assistant. He also selected the brightest, most interesting students in his seminar for longer walks. His "Wunderkinder," he called them.

Although even advanced students and docents stood in awe of Klein, they easily fell into a comradely relationship with Hilbert. His Königsberg accent with its distinctive rhythm and inflection seemed to them to give a unique flavor to everything he said. They delighted in mimicking his manner and opinions, were quick to pick up the "Aber nein!" — *But no!* —with which he announced his fundamental disagreement with an idea, whether in mathematics, economics, philosophy, human relations, or simply the management of the University. ("It was very characteristic the way he said it, but very difficult to catch in English, even in twenty words.")

In the seminar they found him surprisingly attentive to what they had to say. As a rule he corrected them mildly and praised good efforts. But if something seemed too obvious to him he cut it short with "Aber das ist doch ganz einfach!" — *But that is completely simple!* — and when a student made an inadequate presentation he would chastize him or her in a manner that soon became legendary. "Ja, Fräulein S-----, you have given us a very interesting report on a beautiful piece of work, but when I ask myself what have you really said, it is chalk, chalk, nothing but chalk!" And he could also be brutal. "You had better think twice before you uttered a lie or an empty phrase to him," a later student recalled. "His directness could be something to be afraid of."

After a year in Göttingen, the Hilberts decided to build a house on Wilhelm Weber Strasse, the broad linden-lined avenue favored by professors. ("Very likely," wrote Minkowski, "Fate will feel challenged now and try to seduce you from Göttingen with many spectacular offers.") The house was a forthright yellow-brick structure with none of the "new style" ornateness favored by its neighbors. It was large enough that the activities of 4-year-old Franz would not disturb his father as they had in the apartment. The yard in back was large too. They got a dog, the first of a long line of terriers, all to be named Peter. Hilbert, who worked best "under the free sky," hung an 18-foot blackboard from his neighbor's wall and built a covered walk-way so that he could be outdoors even in bad weather.

The house was almost finished when Hilbert wrote the introduction to the *Zahlbericht*. To a later student with a love of language not characteristic of most mathematicians, the introduction was to seem one of the most beautiful parts of German prose, "the style in the literary sense being the accurate image of the way of thinking." In it Hilbert emphasized the esteem in which number theory had always been held by the greatest mathematicians. Even Kronecker was quoted approvingly as "giving expression to the sentiment of his mathematical heart" when he made his famous pronouncement that God made the natural numbers . . . .

"I still find many things to criticize," Minkowski wrote patiently. ". . .Will you not in your foreword perhaps mention the fact that I read the last three sections in manuscript?"

Thus instructed, Hilbert wrote an acknowledgment of what he owed to his friend. Minkowski was still not satisfied.

"That you omitted the thanks to Mrs. Hilbert both Hurwitz and I find scandalous and this simply can't be allowed to remain so."

This last addition was made in the study of the new house at 29 Wilhelm Weber Strasse. The final date on the introduction to the *Zahlbericht* was April 10, 1897.

"I wish you luck that finally after the long years of work the time has arrived when your report will become the common property of all mathematicians," Minkowski wrote upon receiving his specially bound copy, "and I do not doubt in the near future you yourself will be counted among the great classicists of number theory.... Also I congratulate your wife on the good example which she has set for all mathematicians' wives, which now for all time will remain preserved in memory."

The report on algebraic number fields exceeded in every way the expectations of the members of the Mathematical Society. They had asked for a summary of the current state of affairs in the theory. They received a masterpiece, which simply and clearly fitted all the difficult developments of recent times into an elegantly integrated theory. A contemporary reviewer found the *Zahlbericht* an inspired work of art; a later writer called it a veritable jewel of mathematical literature.

The quality of Hilbert's creative contribution in the report is exemplified by that theorem which is still known today simply as "Satz 90." The development of the ideas contained in it were to lead to homological algebra, which plays an important role in algebraic geometry and topology. As another mathematician has remarked, "Hilbert was not only very thorough, but also very fertile for other mathematicians."

For Hilbert, the spring of 1897 was a memorable one — the new house completed, the *Zahlbericht* at last in print. Then came sad news. His only sister, Elise Frenzel, wife of an East Prussian judge, had died in childbirth.

According to a cousin, the relationship between brother and sister was reputed in the family to have been "cool." But for Minkowski, writing to Hilbert at the time, it seemed impossible to find comforting words:

"Whoever knew your sister must have admired her for her always cheerful and pleasant disposition and must have been carried along by her happy approach to life. I still remember... how gay she was in Munich, and how she was in Rauschen. It is really unbelievable that she should have left you so young. How close she must have been to your heart, since you have no other brother or sister and you grew up together for so many years! It seems sometimes that through a preoccupation with science, we acquire a firmer hold over the vicissitudes of life and meet them with greater calm, but in reality we have done no more than to find a way to escape from our sorrows."

Minkowski's next letter, however, contained happy personal news. He had become engaged to Auguste Adler, the daughter of the owner of a leather factory near Strassburg. "My choice is, I am convinced, a happy one and I certainly hope . . . it will be good for my scientific work." In a postscript he added a little information about his fiancée for the Hilberts. "She is 21 years old, she looks very *sympathisch*, not only in my judgment but also in the judgment of all those who know her. She has grown up with six brothers and sisters, is very domestic, and possesses an unusual degree of intelligence."

Minkowski planned to be married in September, but first there was an important event. An International Congress of Mathematicians was going to take place in August in Zürich, which being Swiss was considered appropriately neutral soil. Klein was asked to head the German delegation. "Which will have the consequence," Minkowski noted, "that nobody will come from Berlin."

Although for some reason Hilbert did not attend this first congress, he read the papers which were presented and was most impressed by two of the featured addresses. One of these was a lecture on the modern history of the general theory of functions by Hurwitz. The other was an informal talk by Poincaré on the way in which pure analysis and mathematical physics serve each other.

Shortly after the Congress, Minkowski was married in Strassburg.

He did not write to Hilbert again until the end of November:

"After my long silence, you must think that my marriage has changed me completely. But I stay the same for my friends and for my science. Only I could not show any interest for some time in the usual manner."

With the *Zahlbericht* completed, Hilbert was now involved in investigations of his own which he had long wished to pursue. The focal point of his interest was the generalization of the Law of Reciprocity to algebraic number fields. In classical number theory, the Law of Quadratic Reciprocity, known to Legendre, had been rediscovered by Gauss at the age of 18 and given its first complete proof. Gauss always regarded it as the "gem" of number theory and returned to it five more times during his life to prove it in a different way each time. It describes a beautiful relationship which exists between pairs of primes and the remainders of squares when divided by these.

For treating the Law of Reciprocity in the generality which he had in mind, Hilbert needed a broad foundation; and this he had achieved in the *Zahlbericht*. In its introduction he had noted that "the most richly equipped part of the theory of algebraic number fields appears to me the theory of abelian and relative abelian fields which has been opened up by Kummer for us through his work on the higher reciprocity law and by Kronecker through his investigation of the complex multiplication of elliptic functions. The deep insights into this theory which the works of these two mathematicians give us show at the same time that . . . an abundance of the most precious treasures still lies concealed, beckoning as rich reward to the investigator who knows their value and lovingly practices the art to win them."

Hilbert now proceeded to go after these treasures. As a result of his work on the *Zahlbericht* he had a knowledge of the terrain that was both "intimate and comprehensive." He moved cautiously but with confidence.

"It is a great pleasure," a later mathematician noted, "to watch how, step by step, in a succession of papers ascending from the particular to the general, the adequate concepts and methods are evolved and the essential connections come to light."

By studying the classical Law of Quadratic Reciprocity of Gauss, Hilbert was able to restate it in a simple, elegant way which also applied to algebraic number fields. From this he was then able to guess with brilliant clarity what the reciprocity law must be for degrees higher than 2, although he did not prove his conjectures in all cases. The crown of his work was the paper published the year after the *Zahlbericht* and entitled "On the theory of relative abelian fields." In this paper, which was basically programmatic in character, he sketched out a vast theory of what were to become known as "class-fields," and developed the methods and concepts needed to carry out the necessary investigations. To later mathematicians it was to seem that he had "conceived by divination" — nowhere else in his work is the accuracy of his mathematical intuition so apparent. Unlike the work in invariants, which had marked the end of a development, the work in algebraic number fields was destined to be a beginning. But for other mathematicians.

Hilbert himself now turned abruptly away.

VIII

Tables, Chairs, and Beer Mugs

The announcement that Hilbert would lecture during the winter of 1898–99 on the elements of geometry astonished the students to whom he had talked "only number fields" since his arrival in Göttingen three years before. Yet the new interest was not entirely without antecedent.

In his docent days Hilbert had attended a lecture in Halle by Hermann Wiener on the foundations and structure of geometry. In the station in Berlin on his way back to Königsberg, under the influence of Wiener's abstract point of view in dealing with geometric entities, he had remarked thoughtfully to his companions: "One must be able to say at all times – instead of points, straight lines, and planes – tables, chairs, and beer mugs." In this homely statement lay the essence of the course of lectures which he now planned to present.

To understand Hilbert's approach to geometry, we must remember that in the beginning mathematics was a more or less orderless collection of statements which either seemed self-evident or were obtained in a clear, logical manner from other seemingly self-evident statements. This criterion of evidence was applied without reservation in extending mathematical knowledge. Then, in the third century B. C., a teacher named Euclid organized some of the knowledge of his day in a form that was commonly followed. First he defined the terms he would use – *points*, *lines*, *planes*, and so on. Then he reduced the application of the criterion of evidence to a dozen or so statements the truth of which seemed in general so clear that one could accept them as true without proof. Using only these definitions and axioms (as the statements were later called), he proceeded to derive almost five hundred geometric statements, or theorems. The truth of these was in many cases not at all self-evident, but it was guaranteed by the fact that all the theorems had been derived strictly according to the accepted laws of logic from the definitions and the axioms already accepted as true.

Although Euclid was not the most imaginative of the Greek geometers and the axiomatic method was not original with him, his treatment of geometry was greatly admired. Soon, however, mathematicians began to recognize that in spite of its beauty and perfection there were certain flaws in Euclid's work; particularly, that the axioms were not really sufficient for the derivation of all the theorems. Sometimes other, unstated assumptions crept in — especially assumptions based on visual recognition that in a particular construction certain lines were bound to intersect. It also seemed that one of Euclid's axioms — the Parallel Postulate — went so far beyond the immediate evidence of the senses that it could not really be accepted as true without proof. In its various forms the Parallel Postulate makes a statement essentially equivalent to the statement that through any point not on a given line in a plane, at most one line can be drawn which will not intersect the given line. Generally, however, this flaw and the others in Euclid were dismissed as things which could be easily removed, first by enlarging the original number of axioms to include the unstated assumptions and then by proving the particularly questionable axiom as a theorem, or by replacing it with another more intuitively evident axiom, or — finally — by demonstrating that its negation led to a contradiction. This last and most sophisticated method of dealing with the problem of the Parallel Postulate represents the first appearance in mathematics of the concept of *consistency*, or freedom from contradiction.

Gauss was apparently the first mathematician to whom it occurred, perhaps as early as 1800, that the negation of Euclid's parallel postulate might not lead to a contradiction and that geometries other than Euclid's might be possible. But this idea smacked so of metaphysical speculation that he never published his investigations on the subject and only communicated his thoughts to his closest friends under pledges of secrecy.

During the 1830's, however, two highly individualistic mathematicians tried independently but almost simultaneously to derive from a changed parallel axiom and the other, unchanged traditional axioms of euclidean geometry what theorems they could. Their new axiom stated in essence that through any point not on a given line, infinitely many lines can be drawn which will never meet the given line. Since this was contrary to what they thought they knew as true, the Russian Lobatchewsky and the Hungarian J. Bolyai expected that the application of the axiomatic method would lead to contradictory theorems. Instead, they found that although the theorems established from the new set of axioms were at odds with the results of everyday experience (the angles of a triangle, for instance, did not add up to two right angles as in Euclid's geometry), none of the expected contradictions appeared in the new geometry thus established. It was possible, they had discovered, to build up a consistent geometry upon axioms which (unlike Euclid's) did not seem self-evidently true, or which even appeared false.

Surprisingly enough, this discovery of *non*-euclidean geometries did not arouse the "clamors of the Boeotians" which, according to Gauss (in a letter to Bessel on January 27, 1829), had deterred him from publishing his own investigations on the subject. In fact, there was not very much interest in the discovery among mathematicians. For the majority it seems to have been *too* abstract.

It was not until 1870 that the idea was generally accepted. At that time the 21-year-old Felix Klein discovered a "model" in the work of Cayley by means of which he was able to identify the primitive objects and relations of non-euclidean geometry with certain objects and relations of euclidean geometry. In this way he established that non-euclidean geometry is every bit as consistent as euclidean geometry itself; for a contradiction existing in the one will have of necessity to appear in the other.

Thus the impossibility of demonstrating the Parallel Postulate was at last shown to be "as absolutely certain as any mathematical truth whatsoever." But, again, the full impact of the discovery was not immediately and generally felt. The majority of mathematicians, although they now recognized the several non-euclidean geometries resulting from various changes in the Parallel Postulate, held back from recognizing the fact, which automatically followed, that Euclid's other axioms were equally arbitrary hypotheses for which other hypotheses could be substituted and that still other non-euclidean geometries were possible.

A few mathematicians did try to achieve treatments of geometry which would throw into relief the full implication of the discovery of the non-euclidean geometries, and would at the same time eliminate all the hidden assumptions which had marred the logical beauty of Euclid's work. Such a treatment had been first achieved by Moritz Pasch, who had avoided inadvertently depending on assumptions based on visual evidence by reducing geometry to a pure exercise in logical syntax. Giuseppe Peano had gone even farther. In essence he had translated Pasch's work into the notation of a symbolic logic which he himself had invented. Peano's version of geometry was completely abstract — a calculus of relations between variables.

It was difficult to see how Hilbert could hope to go beyond what had already been done in this area of mathematical thought. But now in his lectures he proceeded to reverse the trend toward absolutely abstract symbolization of geometry in order to reveal its essential nature. He returned to Euclid's points, straight lines and planes and to the old relations of incidence, order and congruence of segments and angles, the familiar figures. But his return did not signify a return to the old deception of euclidean geometry as a statement of truths about the physical universe. Instead — within the classical framework — he attempted to present the modern point of view with even greater clarity than either Pasch or Peano.

With the sure economy of the straight line on the plane, he followed to its logical conclusion the remark which he had made half a dozen years before in the Berlin station. He began by explaining to his audience that Euclid's definitions of point, straight line and plane were really mathematically insignificant. They would come into focus only by their connection with whatever axioms were chosen. In other words, whether they were *called* points, straight lines, planes or were *called* tables, chairs, beer mugs, they would *be* those objects for which the relationships expressed by the axioms were true. In a way this was rather like saying that the meaning of an unknown word becomes increasingly clear as it appears in various contexts. Each additional statement in which it is used eliminates certain of the meanings which would have been true, or meaningful, for the previous statements.

In his lectures Hilbert simply *chose* to use the traditional language of Euclid:

"Let us conceive three distinct systems of things," he said. "The things composing the first system we will call *points* and designate them by the letters A, B, C, ..."

The "things" of the other two systems he called *straight lines* and *planes*. These "things" could have among themselves certain mutual relations which, again, he chose to indicate by such familiar terms as *are situated*, *between*, *parallel*, *congruent*, *continuous*, and so on. But, as with the "things" of the three systems, the meaning of these expressions was not to be determined by one's ordinary experience of them. For example, the primitive terms could denote any objects whatsoever provided that to every pair of objects called *points* there would correspond one and only one of the objects called *straight lines*, and similarly for the other axioms.

The result of this kind of treatment is that the theorems hold true for any interpretation of the primitive notions and fundamental relationships for which the axioms are satisfied. (Many years later Hilbert was absolutely delighted to discover that from the application of a certain set of axioms the laws governing the inheritance of characteristics in the fruit fly can be derived: "So simple and precise and at the same time so miraculous that no daring fantasy could have imagined it!")

In his lectures Hilbert now proposed to set up on this foundation a simple and complete set of independent axioms by means of which it would be possible to prove all the long familiar theorems of Euclid's traditional geometry. His approach — the original combination of the abstract point of view and the concrete traditional language — was peculiarly effective. "It was as if over a landscape wherein but a few men with a superb sense of orientation had found their way in murky twilight, the sun had risen all at once," one of his later students wrote. By developing a set of axioms for euclidean geometry which did not depart too greatly from the spirit of Euclid's own axioms, and by employing a minimum of symbolism, Hilbert was able to present more clearly and more convincingly than either Pasch or Peano the new conception of the nature of the axiomatic method. His approach could be followed by the students in his class who knew only the original *Elements* of Euclid. For established mathematicians, whose first introduction to real mathematics had invariably been the *Elements*, it was particularly attractive, "as if one looked into a face thoroughly familiar and yet sublimely transfigured."

At the time of these lectures on geometry arrangements were being made in Göttingen for the dedication of a monument to Gauss and Wilhelm Weber, the two men — one in mathematics, the other in physics — from whom the University's twofold scientific tradition stemmed. To Klein the dedication ceremony seemed to offer an opportunity to emphasize once again the organic unity of mathematics and the physical sciences. Gauss's observatory had been no ivory tower. In addition to his mathematical discoveries, he had made almost equally important contributions to physics, astronomy, geodesy, electromagnetism, and mechanics. The broadness of his interest had been reinforced by a collaboration with Wilhelm Weber. The two men had invented an electromagnetic telegraph which transmitted over a distance of more than 9,000 feet; the monument was to show the two of them examining this invention. Carrying on and extending the tradition of mathematical abstraction combined with deep interest in physical problems was central to Klein's dream for Göttingen. So now he asked Emil Wiechert to edit his recent lectures on the foundations of electrodynamics for a celebratory volume, and asked Hilbert to do the same for his lectures on the foundations of geometry. (This was the same Wiechert who had been Hilbert's official opponent for his promotion exercises at Königsberg, now also a professor at Göttingen.)

For the published work, as a graceful tribute to Kant, whose *a priori* view of the nature of the geometrical axioms had been discredited by the new view of the axiomatic method, Hilbert chose as his epigraph a quotation from his fellow townsman:

"All human knowledge begins with intuitions, then passes to concepts, and ends with ideas."

Time was short, but he took time to send the proof-sheets of the work to Zürich so that Minkowski could go over them. As always, Minkowski was appreciative and prophetic. The work was, in his opinion, a classic and would have much influence on the thinking of present and future mathematicians.

"It is really not noticeable that you had to work so fast at the end," he assured Hilbert. "Perhaps if you had had more time, it would have lost the quality of freshness."

Minkowski was not too happy in Switzerland. "An open word – take the surprise easy – I would love to go back to Germany." His style of thinking and lecturing was not popular in Zürich "where the students, even the most capable among them, ... are accustomed to get everything spoon-fed." But he hesitated to let his availability be known in Germany. "I feel that even if I had some hope of getting a position, I would still make myself ridiculous in the eyes of many."

Hilbert tried to cheer him up by inviting him to Göttingen for the dedication ceremonies of the Gauss-Weber monument. The days spent there seemed "like a dream" to Minkowski when at the end of a week he had to return to the "hard reality" of Zürich. "But their existence cannot be denied any more than your $18 = 17 + 1$ axiom of arithmetic .... No one who has been in Göttingen recently can fail to be impressed by the stimulating society there."

As soon as Hilbert's lectures, entitled in English *The Foundations of Geometry*, appeared in print, they attracted attention all over the mathematical world.

A German reviewer found the book so beautifully simple that he rashly predicted it would soon be used as a text in elementary instruction.

Poincaré gave his opinion that the work was a classic: "[The contemporary geometers who feel that they have gone to the extreme limit of possible concessions with the non-euclidean geometries based on the negation of the Parallel Postulate] will lose this illusion if they read the work of Professor

Hilbert. In it they will find the barriers behind which they have wished to confine us broken down at every point."

In Poincaré's opinion, the work had but one flaw.

"The logical point of view alone appears to interest Professor Hilbert," he observed. "Being given a sequence of propositions, he finds that all follow logically from the first. With the foundation of this first proposition, with its psychological origin, he does not concern himself.... The axioms are postulated; we do not know from whence they come; it is then as easy to postulate A as C.... His work is thus incomplete, but this is not a criticism I make against him. Incomplete one must indeed resign oneself to be. It is enough that he has made the philosophy of mathematics take a long step forward...."

The American reviewer wrote prophetically, "A widely diffused knowledge of the principles involved will do much for the logical treatment of all science and for clear thinking and writing in general."

The decisive factor in the impact of Hilbert's work, according to Max Dehn, who as a student attended the original lectures, was "the characteristic Hilbertian spirit... combining logical power with intense vitality, disdaining convention and tradition, shaping that which is essential into antitheses with almost Kantian pleasure, taking advantage to the fullest of the freedom of mathematical thought!"

To a large extent, Hilbert, like Euclid himself, had achieved success because of the style and logical perfection of his presentation rather than its originality. But in addition to formulating the modern viewpoint in a way that was attractive and easily grasped, he had done something else which was to be of considerable importance. Having set up in a thoroughly rigorous modern manner the traditional ladder of thought – primitive notions, axioms, theorems – he had proceeded to move on to an entirely new level. In after years, when the approach would have become common, it would be known as metamathematics – literally, "beyond mathematics." For, unlike Euclid, Hilbert required that his axioms satisfy certain logical demands:

That they were *complete*, so that all the theorems could be derived from them.

That they were *independent*, so that the removal of any one axiom from the set would make it impossible to prove at least some of the theorems.

That they were *consistent*, so that no contradictory theorems could be established by reasoning with them.

The most significant aspect of this part of Hilbert's work was the attempted proof of the last requirement — that the axioms be shown to be consistent. This is the equivalent of establishing that reasoning with them will never lead to a contradiction: in short, that they can never be used to prove a statement and at the same time to prove its negation. Under the new conception of a mathematical theory as a system of theorems derived in a deductive way from a set of hypotheses arbitrarily chosen without any restriction as to their truth or meaning, this notion of the consistency of the theory was the only substitute for intuitive truth.

As we have seen, a method of establishing such consistency was already in use. By this method it had been established that any inconsistency which exists in non-euclidean geometry must also exist in euclidean geometry. Thus non-euclidean geometry had been shown to be at least as consistent as euclidean geometry.

Hilbert now took the next step, a step which had apparently occurred to no one else, although it was quite obvious. By the use of analytic geometry, he showed that any contradiction which exists in euclidean geometry must also appear as a contradiction in the arithmetic of real numbers. Both non-euclidean geometry and euclidean geometry were thus shown to be at least as consistent as the arithmetic of real numbers, which was accepted as consistent by all mathematicians.

Within a few months after its publication, Hilbert's little book on the foundations of geometry was a mathematical best seller. Plans were being made to translate it into French and English; it was later translated into other languages. Hilbert's students, who the year before had heard him talk "only algebraic number fields," watched the success of the book in amazement. How had Hilbert been able, once again, to enter a new area of mathematics and produce in it, immediately, great mature work? But even as they asked the question, Hilbert was beginning to publish in still another, entirely new area of mathematics.

IX

Problems

"Pure mathematics grows when old problems are worked out by means of new methods," Klein liked to tell his students. "As a better understanding is thus gained of the older questions, new problems naturally arise."

There is perhaps no better illustration of this statement of Klein's than the project which Hilbert now undertook. In the summer of 1899, immediately after the publication of *The Foundations of Geometry*, he turned to an old and celebrated problem known as the Dirichlet Principle, which involved all the greatest names in the mathematical tradition of Göttingen.

At the heart of this problem was a logical point which had been generally ignored up to the time of Weierstrass. Gauss, Dirichlet, Riemann and others had assumed that, in the case of what is known as the boundary value problem of the Laplace equation, a solution always exists. This assumption was intuitively plausible because in the corresponding physical situation described by the mathematical problem there has to be a definite physical result, or solution. Furthermore, from the mathematical side, Gauss had noted that the boundary value problem of this same equation can be reduced to the problem of minimizing a certain double integral for functions with continuous partial derivatives having the prescribed boundary values. Because of the positive character of this double integral, there is clearly a greatest lower bound for the value of the integral; from this it was further assumed that for one of the functions under consideration the integral would actually have the value of the greatest lower bound.

This mode of reasoning became known as the Dirichlet Principle when Bernhard Riemann used it very freely in his doctoral dissertation in 1851 as the foundation of his geometric function theory and named it in honor of his teacher, Lejeune Dirichlet, who had lectured on the principle in a less general form.

In retrospect, Riemann's dissertation is seen as one of the most important events of modern mathematical history. In its own day, however, it fell into disgrace when Weierstrass objected to the Dirichlet Principle. It was not legitimate mathematically, Weierstrass pointed out, to assume without proof that among the admissible functions there would be one for which the integral would actually have the minimum value.

To someone who is not a mathematician it may seem that this demand of Weierstrass's for a mathematical proof of a principle which obviously works in the physical situations to which it is applied is unreasonable. But this is not so, as Riemann himself recognized when Weierstrass made his criticism. Only a rigorous mathematical proof can establish the ultimate trustworthiness of a mathematical structure and insure that the mathematical description of the physical phenomena to which it is applied is always meaningful.

Riemann was not seriously disturbed by Weierstrass's criticism, however. He had made many of his function-theoretical discoveries on the basis of analogous physical situations, particularly in connection with the behavior of electric currents, and he believed that a problem which made "sensible physics" would make "sensible mathematics." He was confident that the existence of the desired minimum could be established by mathematical proof when and if necessary. But he died young, not yet forty; and a few years after his death, Weierstrass was able to show with finality that the Dirichlet Principle does not in fact invariably hold. He did this by constructing an example for which there is no suitable function that minimizes the integral under the prescribed boundary conditions.

This should have been the end of the Dirichlet Principle, but it was not. Although Riemann's theory was neglected for a while, it was simply too useful in mathematical physics to be discarded. Since the Principle itself was not in general valid, mathematicians invented various ingenious *ad hoc* methods of proving the existence theorems which Riemann had based on the foundation of the Dirichlet Principle. They thus managed to achieve essentially the same end results that he had achieved, but not with the same elegance.

By the time that Hilbert turned his attention to the Dirichlet Principle, mathematicians had given up all hope of salvaging the Principle itself. Only recently Carl Neumann (the son of Franz Neumann), who had done some of the most important work on the subject, had mourned that the Dirichlet Principle, "which is so beautiful and could be utilized so much in the future, has forever sunk from sight."

Unlike many of his contemporaries, who found the demands of rigor a burden, Hilbert firmly believed that rigor contributes to simplicity. He had profound admiration for the way in which Weierstrass had transformed the intuitive analysis of continuity into a strict and logical system. But he refused to let himself be put off by Weierstrass's critique of the Dirichlet Principle. For him, as he said, its "alluring simplicity and indisputable abundance of possible applications" were combined with "a conviction of truth inherent in it."

It was characteristic of Hilbert's mathematical approach to go back to questions in their original conceptual simplicity, and this is what he now did – with, as one of his later pupils commented, "all the naiveté and the freedom from bias and tradition which is characteristic only of truly great investigators." In September 1899, almost fifty years after Riemann's dissertation, he was able to present to the German Mathematical Society a first attempt at what he called, in reference to Neumann's remark, the "resuscitation" of the Dirichlet Principle.

In a few minutes – the whole paper, including the introduction, was scarcely five pages in length – he showed how by placing certain limitations on the nature of the curves and boundary values he could remove Weierstrass's objections and so return Riemann's theory virtually to its original beauty and simplicity. This treatment of the famous problem excited "universal surprise and admiration," according to an American who was present at the meeting. The thought process was simple, but no longer intuitive in any way. Klein commented admiringly: "Hilbert has clipped the hair of the surfaces."

(Half a dozen years later, on the occasion of the 150th anniversary of the Göttingen Scientific Society, he was to return to the Dirichlet Principle and produce a second proof.)

"Hilbert's works on this subject belong to his deepest and most powerful achievements. They are more than the conclusion of a development," a later pupil who also did important work in this same field has written. "Not only was Hilbert's existence proof essentially simplified and generalized through the efforts of many mathematicians; but it was also given an important constructive turn. The physicist Walther Ritz, stimulated by Hilbert, invented from the rehabilitated Dirichlet Principle a powerful method for solving boundary value problems numerically by means of partial differential equations, a method which, just in our time, has shaped the computer into an increasingly succesful tool of numerical mathematics. . . ."

After his success with the Dirichlet Principle, Hilbert decided that during the winter semester of 1899–1900 he would lecture – for the first time in his career — on the calculus of variations. This is the branch of analysis which deals with the type of extremum problems in which (as in the case of Dirichlet's problem) the variable for which a minimizing or maximizing value is sought is not a single numerical variable or a finite number of such variables, but a whole variable curve or function, or even a system of variable functions.

It was a matter of happy experience with Hilbert that great individual problems are the life blood of mathematics. For this reason the calculus of variations had a special charm for him. It was a mathematical theory which had developed from the solution of a single problem.

The problem of "the line of quickest descent" was proposed by Johann Bernoulli at the end of the seventeenth century as a challenge to the mathematicians of his time — and especially to his older brother Jakob, whom he had publicly derided as incompetent. Several people (including Newton) produced solutions; but the "rather inelegant" solution of the scorned older brother surpassed them all. For in it, he recognized what the others had not — that the problem of selecting from an infinity of possible curves the one having a given maximum or minimum property was essentially a new type of problem, demanding for its solution the invention of new methods.

A student who attended Hilbert's lectures on the calculus of variations at this time was Max von Laue.

"...the decisive impression," von Laue wrote of his student days, "was my astonishment at seeing how much information about nature can be obtained by the mathematical method. Profoundest reverence for theory would overcome me when it cast unexpected light on previously obscure facts. Pure mathematics, too, did not fail to impress me, especially in the brilliant courses of David Hilbert."

"This man," the future Nobel Prize winner added, "lives in my memory as perhaps the greatest genius I ever laid eyes on."

The calculus of variations had been made much more rigorous by Weierstrass, but it was still a relatively neglected branch of mathematics. During the winter of his lectures, Hilbert made several important contributions. These included a theorem in which he stated and proved the differentiability conditions of a minimizing arc which assures in many cases the existence of a minimum.

Essentially, however, Hilbert's mathematical interests at this time were more varied than they had been since his docent days in Königsberg. The investigations in geometry continued with a number of papers being published. There was also a paper entitled "The number concept," in which, stimulated by his new-found enthusiasm for the axiomatic method, he proposed that an axiomatic treatment be substituted for the usual "genetic" (as he called it) treatment of the real numbers and introduced the conception of a maximal (or non-extensable) model with his completeness axiom. It was in the midst of this uncharacteristically diversified activity that an invitation arrived for him to make one of the major addresses at the second International Congress of Mathematicians in Paris in the summer of 1900.

The new century seemed to stretch out before him as invitingly as a blank sheet of paper and a freshly sharpened pencil. He would like to make a speech which would be appropriate to the significance of the occasion. In his New Year's letter to Minkowski, he mentioned receiving the invitation and recalled the two speeches from the first International Congress which had so impressed him — the scintillating but technical lecture by Hurwitz on the history of the modern theory of functions and the popular discourse of Poincaré on the reciprocal relationship existing between analysis and physics. He had always wanted to reply to Poincaré with a defense of mathematics for its own sake, but he also had another idea. He had frequently reflected upon the importance of individual problems in the development of mathematics. Perhaps he could discuss the direction of mathematics in the coming century in terms of certain important problems on which mathematicians should concentrate their efforts. What was Minkowski's opinion?

Minkowski wrote he would have to give some thought to the matter.

On January 5, 1900, Minkowski wrote again.

"I have re-read Poincaré's lecture . . . and I find that all his statements are expressed in such a mild form that one cannot take exception to them . . . . Since you will be speaking before specialists, I find a lecture like the one by Hurwitz better than a mere chat like that of Poincaré . . . . Actually it depends not so much on the subject as on the presentation. Still, through the framing of the subject you can make twice as many listeners appear . . . .

"Most alluring would be the attempt at a look into the future and a listing of the problems on which mathematicians should try themselves during the coming century. With such a subject you could have people talking about your lecture decades later."

Minkowski did not fail to point out, however, that there were objections to this subject. Hilbert would probably not want to give away his own ideas for solving certain problems. An international audience would not be so interested in a philosophical discussion as a German audience. Prophecy would not come easy.

There was no reply from Hilbert.

On February 25 Minkowski wrote plaintively to Göttingen.

"How does it happen that one hears nothing from you? My last letter contained only the opinion that if you would give a beautiful lecture, then it would be very beautiful. But it is not easy to give good advice."

But Hilbert had not yet made up his mind about the subject of his speech to the Congress.

On March 29 he consulted Hurwitz.

"I must start preparing for a major talk at Paris, and I am hesitating about a subject . . . . The best would be a view into the future. What do you think about the likely direction in which mathematics will develop during the next century? It would be very interesting and instructive to hear your opinion about that."

There is no record of Hurwitz's reply.

Hilbert continued to mull over the future of mathematics in the twentieth century. By June he still had not produced a lecture, and the program for the Congress was mailed without its being listed.

Minkowski was greatly disappointed: "The desire on my part to travel to the Congress is now almost gone."

Then, in the middle of July, came a package of proof-sheets from Hilbert. Here at last was the text of the lecture. Entitled simply "Mathematical Problems," it was to be delivered in Paris and, at almost the same time, published in the *Nachrichten* of the Göttingen Scientific Society.

There was no more talk on Minkowski's part about not going to Paris.

He read the proof-sheets carefully and with interest. In them Hilbert set forth the importance of problems in determining the lines of development in a science, examined the characteristics of great fruitful problems, and listed the requirements for "solution." Then he presented and discussed 23 individual problems, the solution of which, he was confident, would contribute greatly to the advance of mathematics in the coming century.

The first half dozen problems pertained to the foundations of mathematics and had been suggested by what he considered the great achievements of the century just past: the discovery of the non-euclidean geometries and the clarification of the concept of the arithmetic continuum, or real number system. These problems showed strongly the influence of the recent work on the foundations of geometry and his enthusiasm for the power of the axiomatic method. The other problems were special and individual, some old and well known, some new, all chosen, however, from fields of Hilbert's own past, present, and future interest. The final, twenty-third problem was actually a suggestion for the future rather than a problem — that in the coming century mathematicians should pay more attention to what he considered an unjustly neglected subject — the calculus of variations.

Minkowski commented with special enthusiasm about the second problem on Hilbert's list. This was the first statement of what was to become known in twentieth century mathematics as the "consistency problem."

It will be remembered that in Hilbert's work on the foundations of geometry, he had established the consistency of the geometric axioms by showing that geometry is at least as consistent as the arithmetic of real numbers, which was accepted as consistent by all mathematicians. But what about arithmetic? Is it actually consistent? If arithmetic were to be set up as an axiomatic theory, as Hilbert had suggested in his recent paper on "The number concept," then this question had to be answered.

A lawyer might be satisfied that the "preponderance of evidence" indicated "beyond a reasonable doubt" that arithmetic is indeed free from contradiction. But Hilbert had not chosen to be a lawyer. For him as a mathematician the consistency of arithmetic would have to be established with a degree of certainty which is inconceivable in the law or any other human endeavor except mathematics. In his second problem he had asked for *a mathematical proof* of the consistency of the axioms of the arithmetic of real numbers.

To show the significance of this problem, he had added the following observation:

"If contradictory attributes be assigned to a concept, I say that *mathematically the concept does not exist* . . . . In the case before us, where we are concerned with the axioms of the real numbers in arithmetic, the proof of the consistency of the axioms is at the same time the proof of the mathematical existence of the complete system of real numbers or of the continuum. Indeed, when the proof for the consistency of the axioms shall be fully accomplished, the doubts which have been expressed occasionally as to the existence of the complete system of real numbers will become totally groundless."

This, Hilbert felt, would provide — at last — the answer to Kronecker.

"Highly original it is," Minkowski observed, "to set out as a problem for the future what mathematicians have for the longest time believed themselves already completely to possess — the arithmetic axioms. What will the large number of laymen in the audience say to that? Will their respect for us increase? You will have a fight on your hands with the philosophers too!"

During the next few weeks, Minkowski and Hurwitz studied the proof-sheets of Hilbert's lecture and made suggestions about its presentation to the Congress. They were both concerned that it was overly long. Hilbert had concluded the extensive introduction to his problems with a stirring reiteration of his conviction ("which every mathematician shares, but which no one has as yet supported by a proof") that every definite mathematical problem must necessarily be susceptible of an exact settlement, either in the form of an actual answer to the question asked, or by the proof of the impossibility of its solution and therewith the necessary failure of all attempts. He had then taken the opportunity to deny publicly and emphatically the "Ignoramus et ignorabimus" — *we are ignorant and we shall remain ignorant* — which the writings of Emil duBois-Reymond had made popular during the century which was passing:

"We hear within us the perpetual call. There is the problem. Seek its solution. You can find it by pure reason, for in mathematics there is no *ignorabimus*."

Both Minkowski and Hurwitz thought that this statement would be an effective conclusion to the speech. Then, perhaps, the list of problems itself could be distributed separately to the delegates.

"It is better," Minkowski admonished, "if you don't need the given time completely."

On July 28 Minkowski mailed the corrected proof-sheets back to Hilbert:

"Actually I believe that through this lecture, which indeed every mathematician in the world without exception will be sure to read, your attractiveness for young mathematicians will increase — if that is possible!"

On Sunday, August 5, the two friends met in Paris.

A thousand mathematicians had earlier signified their intention of coming to the Congress and bringing with them almost seven hundred members of their families to visit the Centennial Exhibition; but apparently rumors of crowds, high prices and hot weather had deterred them. On the morning of August 6, when the opening session of the Congress was called to order by Poincaré, the total attendance scarcely exceeded 250.

After the opening day, the mathematicians left the alien grounds of the Exhibition and retired to the hill on the left bank where the École Polytechnique and the École Normale Supérieure flank Napoleon's Panthéon and a narrow street leads down to the dingy buildings of the Sorbonne.

Although the past century had seen the development of many new branches of mathematics, the classification for the meetings of the Congress remained traditional. Pure mathematics was represented by sections on

Arithmetic, Algebra, Geometry and Analysis; applied mathematics, by a section on Mechanics. The general sections, which dealt with Bibliography and History in one and Teaching and Methods in the other, were considered of inferior rank to the mathematical sections, the lectures being of more general interest but "not necessarily the most valuable mathematically," according to an American who was there. Originally, Hilbert's talk had been planned for the opening session, but because of his tardiness it had to be delivered at the joint session of the two general sections on the morning of Wednesday, August 8.

The man who came to the rostrum that morning was not quite forty, of middle height and build, wiry, quick, with a noticeably high forehead, bald except for wisps of still reddish hair. Glasses were set firmly on a strong nose. There was a small beard, a still somewhat straggly moustache and under it a mouth surprisingly wide and generous for the delicate chin. Bright blue eyes looked innocently but firmly out from behind shining lenses. In spite of the generally unpretentious appearance of the speaker, there was about him a striking quality of intensity and intelligence.

He had already had an extract of his speech in French distributed. This was not at that time a common practice, and the members of his audience were surprised and grateful.

Slowly and carefully for those who did not understand German well, he began to speak.

X

The Future of Mathematics[1]

"Who of us would not be glad to lift the veil behind which the future lies hidden; to cast a glance at the next advances of our science and at the secrets of its development during future centuries? What particular goals will there be toward which the leading mathematical spirits of coming generations will strive? What new methods and new facts will the new centuries disclose in the wide and rich field of mathematical thought?

"History teaches the continuity of the development of science. We know that every age has its own problems, which the following age either solves or casts aside as profitless and replaces by new ones. If we would obtain an idea of the probable development of mathematical knowledge in the immediate future, we must let the unsettled questions pass before our minds and look over the problems which the science of today sets and whose solution we expect from the future. To such a review of problems the present day, lying at the meeting of the centuries, seems to me well adapted. For the close of a great epoch not only invites us to look back into the past but also directs our thoughts to the unknown future.

"The deep significance of certain problems for the advance of mathematical science in general and the important role which they play in the work of the individual investigator are not to be denied. As long as a branch of science offers an abundance of problems, so long is it alive: a lack of problems foreshadows extinction or the cessation of independent

---

[1] The general remarks from the talk on "Mathematical Problems," delivered by David Hilbert before the Second International Congress of Mathematicians at Paris in 1900, are reprinted with permission of the publisher, the American Mathematical Society, from the *Bulletin of the American Mathematical Society*, vol. 8, 1902, pp. 437–445, 478–479. Copyright 1902. The entire text of the talk appears in the *Bulletin* translated into English by Dr. Mary Winston Newson with the approval of Prof. Hilbert.

development. Just as every human undertaking pursues certain objectives, so also mathematical research requires its problems. It is by the solution of problems that the strength of the investigator is hardened; he finds new methods and new outlooks, and gains a wider and freer horizon.

"It is difficult and often impossible to judge the value of a problem correctly in advance; for the final award depends upon the gain which science obtains from the problem. Nevertheless, we can ask whether there are general criteria which mark a good mathematical problem. An old French mathematician said: 'A mathematical theory is not to be considered complete until you have made it so clear that you can explain it to the first man whom you meet on the street.' This clarity and ease of comprehension, here insisted on for a mathematical theory, I should still more demand for a mathematical problem if it is to be perfect; for what is clear and easily comprehended attracts, the complicated repels us.

"Moreover a mathematical problem should be difficult in order to entice us, yet not completely inaccessible, lest it mock our efforts. It should be to us a guidepost on the tortuous paths to hidden truths, ultimately rewarding us by the pleasure in the successful solution.

"The mathematicians of past centuries were accustomed to devote themselves to the solution of difficult individual problems with passionate zeal. They knew the value of difficult problems. I remind you only of the 'problem of the line of quickest descent,' proposed by Johann Bernoulli. Experience teaches, Bernoulli explained in the public announcement of this problem, that lofty minds are led to strive for the advance of science by nothing more than laying before them difficult and at the same time useful problems, and he therefore hoped to earn the thanks of the mathematical world by following the example of men like Mersenne, Pascal, Fermat, Viviani and others in laying before the distinguished analysts of his time a problem by which, as a touchstone, they might test the value of their methods and measure their strength. The calculus of variations owes its origin to this problem of Bernoulli's and to similar problems.

"Fermat has asserted, as is well known, that the diophantine equation $x^n + y^n = z^n$ (x, y and z integers) is unsolvable — except in certain self-evident cases. The attempt to prove this impossibility offers a striking example of the inspiring effect which such a very special and apparently unimportant problem may have upon science. For Kummer, spurred on by Fermat's problem, was led to the introduction of ideal numbers and to the discovery of the law of the unique decomposition of the numbers of a cyclotomic field into ideal prime factors — a law which today, in its generalization to any algebraic field by Dedekind and Kronecker, stands at the center of the modern theory of numbers and the significance of which extends far beyond the boundaries of number theory and into the realm of algebra and the theory of functions.

"To speak of a very different region of research, I remind you of the problem of the three bodies. The fruitful methods and the far-reaching principles which Poincaré has brought into celestial mechanics and which are today recognized and applied in practical astronomy are due to the circumstance that he undertook to treat anew that difficult problem and to come nearer a solution.

"The two last mentioned problems – that of Fermat and the problem of the three bodies – seem to us almost like opposite poles – the former a free invention of pure reason, belonging to the region of abstract number theory, the latter forced upon us by astronomy and necessary to an understanding of the simplest fundamental phenomena of nature.

"But it often happens also that the same special problem finds application in the most unlike branches of mathematical knowledge. So, for example, the problem of the shortest line plays a chief and historically important part in the foundations of geometry, in the theory of lines and surfaces, in mechanics and in the calculus of variations. And how convincingly has F. Klein, in his work on the icosahedron, pictured the significance of the problem of the regular polyhedra in elementary geometry, in group theory, in the theory of equations, and in that of linear differential equations.

"In order to throw light on the importance of certain problems, I may also refer to Weierstrass, who spoke of it as his happy fortune that he found at the outset of his scientific career a problem so important as Jacobi's problem of inversion on which to work.

"Having now recalled to mind the general importance of problems in mathematics, let us turn to the question of the sources from which this science derives its problems. Surely the first and oldest problems in every branch of mathematics stem from experience and are suggested by the world of external phenomena. Even the rules of calculation with integers must have been discovered in this fashion in a lower stage of human civilization, just as the child of today learns the application of these laws by empirical methods. The same is true of the first problems of geometry, the problems bequeathed to us by antiquity, such as the duplication of the cube, the squaring of the circle; also the oldest problems in the theory of the solution of numerical equations, in the theory of curves and the differential and integral calculus, in the calculus of variations, the theory of

Fourier series, and the theory of potential — to say nothing of the further abundance of problems properly belonging to mechanics, astronomy and physics.

"But, in the further development of a branch of mathematics, the human mind, encouraged by the success of its solutions, becomes conscious of its independence. By means of logical combination, generalization, specialization, by separating and collecting ideas in fortunate ways — often without appreciable influence from without — it evolves from itself alone new and fruitful problems, and appears then itself as the real questioner. Thus arose the problem of prime numers and the other problems of number theory, Galois's theory of equations, the theory of algebraic invariants, the theory of abelian and automorphic functions; indeed almost all the nicer questions of modern arithmetic and function theory arise in this way.

"In the meantime, while the creative power of pure reason is at work, the outer world again comes into play, forces upon us new questions from actual experience, opens up new branches of mathematics; and while we seek to conquer these new fields of knowledge for the realm of pure thought, we often find the answers to old unsolved problems and thus at the same time advance most successfully the old theories. And it seems to me that the numerous and surprising analogies and that apparently pre-established harmony which the mathematician so often perceives in the questions, methods and ideas of the various branches of his science, have their origin in this ever-recurring interplay between thought and experience.

"It remains to discuss briefly what general requirements may be justly laid down for the solution of a mathematical problem. I should say first of all, this: that it shall be possible to establish the correctness of the solution by means of a finite number of steps based upon a finite number of hypotheses which are implied in the statement of the problem and which must be exactly formulated. This requirement of logical deduction by means of a finite number of processes is simply the requirement of rigor in reasoning. Indeed the requirement of rigor, which has become a byword in mathematics, corresponds to a universal philosophical necessity of our understanding; on the other hand, only by satisfying this requirement do the thought content and the suggestiveness of the problem attain their full effect. A new problem, especially when it comes from the outer world of experience, is like a young twig, which thrives and bears fruit only when it is grafted carefully and in accordance with strict horticultural rules upon the old stem, the established achievements of our mathematical science.

"It is an error to believe that rigor in the proof is the enemy of simplicity. On the contrary, we find it confirmed by numerous examples that the rigorous method is at the same time the simpler and the more easily comprehended. The very effort for rigor forces us to discover simpler methods of proof. It also frequently leads the way to methods which are more capable of development than the old methods of less rigor. Thus the theory of algebraic curves experienced a considerable simplification and attained a greater unity by means of the more rigorous function-theoretical methods and the consistent introduction of transcendental devices. Further, the proof that the power series permits the application of the four elementary arithmetical operations as well as the term by term differentiation and integration, and the recognition of the utility of the power series depending upon this proof contributed materially to the simplification of all analysis, particularly of the theory of elimination and the theory of differential equations, and also of the existence proofs demanded in those theories. But the most striking example of my statement is the calculus of variations. The treatment of the first and second variations of definite integrals required in part extremely complicated calculations, and the processes applied by the old mathematicians had not the needful rigor. Weierstrass showed us the way to a new and sure foundation of the calculus of variations. By the examples of the simple and double integral I will show briefly, at the close of my lecture, how this way leads at once to a surprising simplification of the calculus of variations. For in the demonstration of the necessary and sufficient criteria for the occurrence of a maximum and minimum, the calculation of the second variation and in part, indeed, the tiresome reasoning connected with the first variation may be completely dispensed with — to say nothing of the advance which is involved in the removal of the restriction to variations for which the differential coefficients of the function vary but slightly.

"While insisting on rigor in the proof as a requirement for a perfect solution of a problem, I should like, on the other hand, to oppose the opinion that only the concepts of analysis, or even those of arithmetic alone, are susceptible of a fully rigorous treatment. This opinion, occasionally advocated by eminent men, I consider entirely erroneous. Such a one-sided interpretation of the requirement of rigor would soon lead to the ignoring of all concepts arising from geometry, mechanics and physics, to a stoppage of the flow of new material from the outside world, and finally, indeed, as a last consequence, to the rejection of the ideas of the continuum and of the irrational number. But what an important nerve, vital to mathematical science, would be cut by rooting out geometry and mathematical physics! On the contrary, I think that wherever mathematical ideas come up, whether from the side of the theory of knowledge or in geometry, or from the theories of natural or physical science, the problem arises for mathematics to investigate the principles underlying these ideas and so to establish them upon a simple and complete system of axioms, that the exactness of the new ideas and their applicability to deduction shall be in no respect inferior to those of the old arithmetical concepts.

"To new concepts correspond, necessarily, new symbols. These we choose in such a way that they remind us of the phenomena which were the occasion for the formation of the new concepts. So the geometrical figures are signs or mnemonic symbols of space intuition and are used as such by all mathematicians. Who does not always use along with the double inequality $a > b > c$ the picture of three points following one another on a straight line as the geometrical picture of the idea 'between'? Who does not make use of drawings of segments and rectangles enclosed in one another, when it is required to prove with perfect rigor a difficult theorem on the continuity of functions or the existence of points of condensation? Who could dispense with the figure of the triangle, the circle with its center, or with the cross of three perpendicular axes? Or who would give up the representation of the vector field, or the picture of a family of curves or surfaces with its envelope which plays so important a part in differential geometry, in the theory of differential equations, in the foundations of the calculus of variations, and in other purely mathematical sciences?

"The arithmetical symbols are written figures and the geometrical figures are drawn formulas; and no mathematician could spare these drawn formulas, any more than in calculation he could dispense with the insertion and removal of parentheses or the use of other analytical signs.

"The use of geometrical symbols as a means of strict proof presupposes the exact knowledge and complete mastery of the axioms which lie at the foundation of those figures; and in order that these geometrical figures may be incorporated in the general treasure of mathematical symbols, a rigorous axiomatic investigation of their conceptual content is necessary. Just as in adding two numbers, one must place the digits under each other in the right order so that only the rules of calculation, i.e., the axioms of arithmetic, determine the correct use of the digits, so the use of geometrical symbols is determined by the axioms of geometrical concepts and their combinations.

"The agreement between geometrical and arithmetical thought is shown also in that we do not habitually follow the chain of reasoning back to the axioms in arithmetical discussions, any more than in geometrical. On the contrary, especially in first attacking a problem, we apply a rapid, unconscious, not absolutely sure combination, trusting to a certain arithmetical feeling for the behavior of the arithmetical symbols, which we could dispense with as little in arithmetic as with the geometrical imagination in geometry. As an example of an arithmetical theory operating rigorously with geometrical ideas and symbols, I may mention Minkowski's work, *The Geometry of Numbers*.

"Some remarks upon the difficulties which mathematical problems may offer, and the means of overcoming them, may be in place here.

"If we do not succeed in solving a mathematical problem, the reason frequently consists in our failure to recognize the more general standpoint from which the problem before us appears only as a single link in a chain of related problems. After finding this standpoint, not only is this problem frequently more accessible to our investigation, but at the same time we come into possession of a method which is applicable also to related problems. The introduction of complex paths of integration by Cauchy and of the notion of the ideals in number theory by Kummer may serve as examples. This way for finding general methods is certainly the most practical and the most certain; for he who seeks for methods without having a definite problem in mind seeks for the most part in vain.

"In dealing with mathematical problems, specialization plays, as I believe, a still more important part than generalization. Perhaps in most cases where we unsuccessfully seek the answer to a question, the cause of the failure lies in the fact that problems simpler and easier than the one in hand have been either incompletely solved, or not solved at all. Everything depends, then, on finding those easier problems and on solving them by means of devices as perfect as possible and of concepts capable of generalization. This rule is one of the most important levers for overcoming mathematical difficulties; and it seems to me that it is used almost always, though perhaps unconsciously.

"Occasionally it happens that we seek the solution under insufficient hypotheses or in an incorrect sense, and for this reason do not succeed. The problem then arises: to show the impossibility of the solution under the given hypotheses, or in the sense contemplated. Such proofs of impossibility were effected by the ancients; for instance, when they showed that the ratio of the hypotenuse to the side of an isosceles right triangle is irrational. In later mathematics, the question as to the impossibility of certain solutions plays a prominent part; and we perceive in this way that old and difficult problems, such as the proof of the axiom of parallels, the squaring of the circle, or the solution of equations of the fifth degree by radicals, have finally found fully satisfactory and rigorous solutions, although in another sense from that originally intended. It is probably this remarkable fact along with other philosophical reasons that gives rise to the conviction (which every mathematician shares, but which no one has as yet supported by a proof) that every definite mathematical problem must necessarily be susceptible of an exact settlement, either in the form of an actual answer to the question asked, or by the proof of the impossibility of its solution and therewith the necessary failure of all attempts. Take any definite unsolved problem, such as the question as to the irrationality of the Euler-Mascheroni constant C or the existence of an infinite number of prime numbers of the form $2^n + 1$. However unapproachable these problems may seem to us and however helpless we stand before them, we have, nevertheless, the firm conviction that their solution must follow by a finite number of purely logical processes.

"Is this axiom of the solvability of every problem a peculiarity characteristic only of mathematical thought, or is it possibly a general law inherent in the nature of the mind, a belief that all questions which it asks must be answerable by it? For in other sciences also one meets old problems which have been settled in a manner most satisfactory and most useful to science by the proof of their impossibility. I cite the problem of perpetual motion. After seeking unsuccessfully for the construction of a perpetual motion machine, scientists investigated the relations which must subsist between the forces of nature if such a machine is to be impossible; and this inverted question led to the discovery of the law of the conservation of energy, which, again, explained the impossibility of perpetual motion in the sense originally intended.

"This conviction of the solvability of every mathematical problem is a powerful incentive to the worker. We hear within us the perpetual call: There is the problem. Seek its solution. You can find it by pure reason, for in mathematics there is no *ignorabimus.*"

At this point, in an effort to shorten his talk as Minkowski and Hurwitz had urged, Hilbert presented only 10 problems out of the total of 23 which he had listed in his manuscript. (However, because the problems were to become known by their numbered position in the full list, they are given here numbered in that way.)[2]

---

[2] A summary in French of Hilbert's talk and the list of the problems presented appears in *L'enseignement mathématique*, vol. 2, 1900, pp. 349–355.

The first three problems concerned the foundations of mathematics:

1. To prove Cantor's "continuum hypothesis" that any set of real numbers can be put into one-to-one correspondence either with the set of natural numbers or with the set of all real numbers (i.e., the continuum).
2. To investigate the consistency of the arithmetic axioms.
6. To axiomatize those physical sciences in which mathematics plays an important role.

"So far we have considered only questions concerning the foundations of the mathematical sciences. Indeed, the study of the foundations of a science is always particularly attractive, and the testing of these foundations will always be among the foremost problems of the investigator. Weierstrass said, 'The final objective always to be kept in mind is to arrive at a correct understanding of the foundations . . . . But to make any progress in the sciences the study of individual problems is, of course, indispensable.' In fact, a thorough understanding of its special theories is necessary to the successful treatment of the foundations of the science. Only that architect is in the position to lay a sure foundation for a structure who knows its purpose thoroughly and in detail."

The next four problems were selected from arithmetic and algebra:

7. To establish the transcendence, or at least the irrationality, of certain numbers.
8. To prove the correctness of an extremely important statement by Riemann that the zeros of the function known as the "zeta function" all have the real part $^1/_2$, except the well known negative integral real zeros.
13. To show the impossibility of the solution of the general equation of the 7th degree by means of functions of only two arguments.
16. To conduct a thorough investigation of the relative position of the separate branches which a plane algebraic curve of *n*th order can have when their number is the maximum . . . and the corresponding investigation as to the number, form, and position of the sheets of an algebraic surface in space.

The last three problems came from the theory of functions:

19. To determine whether the solutions of "regular" problems in the calculus of variations are necessarily analytic.
21. To show that there always exists a linear differential equation of the Fuchsian class with given singular points and monodromic group.
22. To generalize a theorem proved by Poincaré to the effect that it is always possible to uniformize any algebraic relation between two variables by the use of automorphic functions of one variable.

"The problems mentioned," Hilbert told his audience, "are merely samples of problems; yet they are sufficient to show how rich, how manifold and how extensive the mathematical science is today; and the question is urged upon us whether mathematics is doomed to the fate of those other sciences that have split up into separate branches, whose representatives scarcely understand one another and whose connection becomes ever more loose. I do not believe this nor wish it. Mathematical science is in my opinion an indivisible whole, an organism whose vitality is conditioned upon the connection of its parts. For with all the variety of mathematical knowledge, we are still clearly conscious of the similarity of the logical devices, the *relationship* of the *ideas* in mathematics as a whole and the numerous analogies in its different departments. We also notice that, the farther a mathematical theory is developed, the more harmoniously and uniformly does its construction proceed, and unexpected relations are disclosed between hitherto separate branches of the science. So it happens that, with the extension of mathematics, its organic character is not lost but only manifests itself the more clearly.

"But, we ask, with the extension of mathematical knowledge will it not finally become impossible for the single investigator to embrace all departments of this knowledge? In answer let me point out how thoroughly it is ingrained in mathematical science that every real advance goes hand in hand with the invention of sharper tools and simpler methods which at the same time assist in understanding earlier theories and cast aside older, more complicated developments. It is therefore possible for the individual investigator, when he masters these sharper tools and simpler methods, to find his way more easily in the various branches of mathematics than is possible in any other science.

"The organic unity of mathematics is inherent in the nature of this science, for mathematics is the foundation of all exact knowledge of natural phenomena. That it may completely fulfill this high destiny, may the new century bring it gifted prophets and many zealous and enthusiastic disciples!"

XI

The New Century

It was hot and sultry in the lecture hall at the Sorbonne by the time that Hilbert finished his lecture on mathematical problems, and the discussion which followed was "somewhat desultory," according to the reporter for the American Mathematical Society.

". . . the claim was made, although apparently without adequate grounds, that more had been done as regards the equation of the 7th degree (by some German writer) than the author of the paper was willing to allow. A more precise objection was taken to M. Hilbert's remarks on the axioms of arithmetic by M. Peano, who claimed that such a system as that specified as desirable had already been established by [some of] his compatriots . . . ."

The real news of the day, as reported in the special edition of The New York Times which was distributed on the grounds of the Exhibition, was that the United States, Great Britain, Germany and Japan were going to have to carry out their military program in China without any more troops from Russia and France, who were otherwise occupied on the Siberian frontier and in Indo-China. The King of Italy had recently been assassinated and the country was in turmoil. Queen Victoria planned to address Parliament. William Jennings Bryan was informed that he had again been chosen to head the Democratic ticket against President McKinley.

But up at the Sorbonne, in the remaining days of the Congress, it became quite clear that David Hilbert had captured the imagination of the mathematical world with his list of problems for the twentieth century. His practical experience seemed to guarantee that they met the criteria which he had set up in his lecture; his judgment, that they could actually be solved in the years to come. His rapidly growing fame — exceeded now only by that of Poincaré — promised that a mathematician could make a reputation for himself by solving one of the Paris problems.

Immediately after the Congress adjourned, Hilbert went for a short vacation at Rauschen. Receiving a note from him, Minkowski recalled the "beautiful times" they used to have upon the strand: "With pleasure I also see what I have of course known for a long time — that one can learn much from you, not only in mathematics, but also in the art of enjoying life sensibly like a philosopher."

The first important result in connection with one of Hilbert's problems came within the year 1900. His own student, 22-year-old Max Dehn, showed that (as Hilbert had conjectured) a regular tetrahedron cannot be cut up and reassembled into a cube of equal volume. This provided a partial solution for the third problem. The next year Dehn completed the solution. He thus became the first mathematician to pass into what was later to be known as "the honors class" of mathematicians who had solved or contributed to the solution of one of David Hilbert's 23 Paris problems.

After Paris, Hilbert himself continued to investigate geometric questions, but for the most part he devoted himself to questions of analysis. This is an area of mathematics that differs in an important respect from those in which he had previously worked. In arithmetic and algebra, calculations ordinarily involve only a finite number of quantities and end after a finite number of steps. Analysis operates in a continuum. Solutions are achieved by showing that infinite series of numbers converge to a limit. By now, Hilbert was the enthusiastic champion of the axiomatic method; and he thought he saw in analysis an opportunity to exercise the impressive powers of this method to unify, order, and clarify.

"It appears to me of outstanding interest," he was later to remark, "to undertake an investigation of the convergence conditions which serve for the erection of a given analytic discipline so that we can set up a system of the simplest fundamental facts which require for their proof a specific convergence condition. Then by the use of that one convergence condition alone — without the addition of any other convergence condition whatsoever — the totality of the theorems of the particular discipline can be established."

This was somewhat similar to what Riemann had attempted to do with the Dirichlet Principle, and Hilbert thought that he also could find "the one simple fundamental fact" which he needed in the calculus of variations. Then, one day in the winter of 1900–1901, a Swedish student brought to Hilbert's seminar a recently published paper on integral equations by a countryman named Ivar Fredholm.

Integral equations are functional equations, the history of which is closely tied up with the problems of mathematical physics, particularly the problem of the oscillations of a continuum. A theory of sorts for this type of equation had been very slowly evolving. But now Fredholm presented the solution of particular equations (which came to be named after him) in an original and elegant way that revealed a suggestive analogy between integral equations and the linear equations of algebra.

Hilbert promptly recognized that in this work Fredholm was coming closer to the desired goal of a unifying methodological approach to analysis than he himself was in his work in the calculus of variations. It was a matter of pride with him that he was not emotionally committed to any program — that he saw things as they were rather than as he would like them to be. So now, without regrets, he turned away from his own program and charged the subject of integral equations with impressive intensity. It would have to remain forever undecided (as Blumenthal observed) whether Hilbert would have been able to bring to the methods of the calculus of variations the flexibility and power necessary to penetrate and sustain the whole of analysis. For now Hilbert talked *only* about integral equations to his students.

It was at this same time that a young Japanese named Teiji Takagi came to Göttingen, his way paid by his country. He was to become one of the half dozen mathematicians who would develop the ideas of class-fields which Hilbert had sketched out in his last paper on algebraic number fields. He had already written a little book on *New Arithmetic*, very simple in comparison with the recent number theory work of Hilbert but much advanced for the mathematical level of his native land at that time; now he was looking forward to studying with the author of the *Zahlbericht*. But when Takagi arrived in Göttingen, Hilbert had nothing at all to say to him on the subject of number theory. Instead, he was already sketching out to his students in conversations and lectures some of the ideas which he would ultimately use in his general theory of integral equations.

Another student who arrived at this time was Erhard Schmidt. He had come down from Berlin to "scout" the mathematical education in Göttingen and compare it with that being offered in the capital by the formidable trio of Fuchs, H. A Schwarz, and Frobenius. Fuchs was the same Fuchs with whom Hilbert had studied at Heidelberg; Schwarz, who had been responsible for bringing Klein to Göttingen, conducted a twice-a-month colloquium that was internationally famous; Frobenius, it was said, delivered the most perfect mathematical lectures in Germany — "their only fault being," according to one student, "that because of their completeness they failed even to suggest the existence of unsolved problems." But young Schmidt was so impressed with the mathematics in Göttingen that he decided not to return to Berlin.

Although the subject of discussion had changed, the weekly seminar walks continued. Now, though, the calm of the countryside was frequently disturbed by the snorting of an engine-powered monster. Hilbert's friend Nernst had bought one of the new motor cars; and the hills that defeated other motorists were no problem for him. He merely turned the tap on the cylinder of N_2O he had installed on the dashboard, thus injecting laughing gas into the mixture, and roared triumphantly up any hill.

During the winter semester of 1901–1902, Hilbert was lecturing on potential theory and applying his first results on integral equations. Because of their newness, his ideas were sometimes almost impossible for his students to follow. Even the Lesezimmer notes, worked out by Albert Andrae, the assistant, were not too helpful. In fact, Andrae sometimes penciled warnings on the notes: "From page such-and-such to page so-and-so, no guarantee of correctness can be given." At the Christmas celebration of the Mathematics Club one befuddled student of potential theory read the following gently ironical lines: "Der eine bleibt erst unverständlich/ Der Andrae macht es klar." (The verse depended on the similarity between the name Andrae and the German word for "other" – *The one is at first obscure, the other makes it clear.*)

By spring Erhard Schmidt's enthusiasm for mathematics at Göttingen had brought some of his friends from Berlin. One of these was Constantin Carathéodory, the scion of a powerful Greek family, who at 26 had given up a promising career as an engineer to return to school and devote himself to the study of pure mathematics. His family considered his plan foolishly romantic: one does not usually begin a mathematical career at 26. "But I could not resist the obsession that through unrestrained preoccupation with mathematics my life would become worthwhile."

By now Hilbert was just about as famous as it is possible for a mathematician to be. He accepted his success, as Otto Blumenthal noted, "with a naive, mild pleasure, not letting himself be confused into false modesty." The string of victories which had been inaugurated by the solution of Gordan's Problem twelve short years ago reminded Blumenthal (now a Privatdozent himself) of Napoleon's Italian campaign: the climactic work in invariant theory – the *Zahlbericht* and the deep, fertile program for class-fields – the widely read and influential little book on the foundations of geometry – the salvaging of the Dirichlet Principle – the important theorems in the calculus of variations – the Paris Problems. Foreign academies elected Hilbert to membership. The German government awarded him the title of Geheimrat, roughly equivalent to an English knighthood.

Someone, trying to help Hilbert with something and addressing him over and over as "Herr Geheimrat," noted that he appeared irritated and asked worriedly, "Am I bothering you, Herr Geheimrat?"

"Nothing about you bothers me," Hilbert shot back, "except your obsequiousness!"

Klein, after becoming a Geheimrat, always insisted on being addressed by that title. How did Hilbert prefer to be addressed?

"Hilbert?" a former student says. "He didn't care. He was a king. He was Hilbert."

The parents were still alive at this time. Judge Hilbert had long remained "suspicious" of his son's profession and his success. Mathematics being what it is, he could never as an outsider really appreciate the quality of the achievement; but he must have been reassured at last by the honors.

Minkowski, on a visit to Göttingen, was much impressed with the mathematical atmosphere around his friend.

"Even through merely a sojourn in such air," he wrote after his return to the still unsatisfactory situation in Zürich, "a person receives an increased desire to do great things . . . . I am already at work on a paper for the *Annalen*."

And yet, as Hilbert passed his fortieth birthday on January 23, 1902, he was not entirely happy.

Although he and Klein shared "a perfect trust and common interest" (his own words), they were not intimate. More and more since Hilbert's coming to Göttingen, Klein had devoted himself to activities which were not in themselves mathematics. In addition to his teaching and administrative duties, he was the moving force behind a number of projects: the plan for the 30-volume mathematical encyclopedia; the newly organized International Schools Commission, which had been set up to study the development of teaching methods "in all civilized countries" from kindergarten to graduate school; the attempt to improve and enlarge the scientific education offered in the German middle schools and to bring together, at the university level, technical and mathematical training; the dream, cherished since his visit to America, of fertilizing technology itself with the methods of pure mathematics. Hilbert had scarcely any interest in these projects of Klein's.

Also, with age, Klein was becoming more olympian. A favorite joke among the students was the following: In Göttingen there are two kinds of mathematicians, those who do what they want and not what Klein wants —

and those who do what Klein wants and not what they want. Klein is not either kind. Therefore, Klein is not a mathematician.

Klein was interested in his students and spent long hours in conference with them, but he remained always a superior being. He dispensed ideas, one of the students later said, "with royal joyousness from his own wealth" and "directed each one with unfailing certainty to the point which best accorded with his individuality." Among themselves, they called him "the great Felix." It was told in Göttingen that at dinner in Klein's house a student was sometimes so awed by his host that he stood up when he was asked a question.

Hilbert did not feel personally threatened by Klein, however. A few years before, when he had been offered Sophus Lie's chair in Leipzig, he had consulted Minkowski about the advisability of his accepting the opportunity to leave Göttingen. Minkowski had pointed out that perhaps if he were "spatially separated" from Klein, "outsiders" might more easily recognize that it was Hilbert who was now the greatest German mathematician. But this argument had no effect. Hilbert had refused the Leipzig offer.

Now, however, he was becoming increasingly aware that in his relationship with Klein something was missing which was necessary to him — something which Klein because of his nature could not provide. Then, a few months after his fortieth birthday, he received another opportunity to leave Göttingen. Lazarus Fuchs died and the honored position — Fuchs's chair in Berlin — was offered to Hilbert.

When the news of Hilbert's "call" became known among the docents and advanced students, they were greatly upset. Many of them had come to Göttingen just because Hilbert was there. Some, like Erhard Schmidt and his friends, had come from Berlin itself. Yet it seemed to all only natural that Hilbert, the leading German mathematician, should want to take his place in the capital. Although they did not have much hope of affecting his decision, they appointed three students, headed by Walther Lietzmann, to go to his house and petition him to remain in Göttingen. Mrs. Hilbert served them punch in the garden. Hilbert listened without comment to what they had to say. They left discouraged. The length of time he was taking in coming to his decision, his frequent trips to the capital, his unusual nervousness in lectures, all led them to believe that he was planning to accept the Berlin offer.

Hilbert, however, was actually attempting to solve a personal problem in the way that he had grown accustomed to solving mathematical problems. He did not want to leave Göttingen. As he had explained to Minkowski at the time of the Leipzig offer, he felt that he had more vigor for his work in a little town, which made scientific interchange easy and provided many opportunities for contact with nature, than he would have in a big city. He was well aware of the benefits he reaped from the administrative genius of Klein. He also had a sense of the appropriateness of the leading German mathematician's being at the university of Gauss. But to stay happily in Göttingen, he knew he needed to have some colleague who would provide the close scientific and personal companionship which at Königsberg he had received from Minkowski. The solution to the problem was obvious. There was but one unwritten rule regarding an offer from another university. In addition to trying to improve one's own professonal situation, one was also expected to try to improve the situation of his department and his subject. Recently, Nernst, offered a position in Munich, had extracted as the price of his remaining in Göttingen the first physico-chemical laboratory in Germany. Now, with Klein's approval and support, Hilbert proposed to Althoff that a new professorship of mathematics be created at Göttingen and that it be offered to Minkowski.

He had not raised Minkowski's hopes of returning to Germany until the new position was definite. Then he saw to it that Minkowski was in Göttingen presenting a paper to the scientific society on the day the announcement was made.

The diplomatic skill with which he conducted this bold and ultimately successful maneuver is glimpsed in the fact that he always gave Althoff the full credit:

"It was again Althoff who transplanted Minkowski to soil better suited to him. With an intrepidity unprecedented in the history of the administration of the Prussian universities, Althoff created here in Göttingen a new professorship . . . ."

When the members of the Mathematics Club heard the news — "that Hilbert was staying and Minkowski was coming" — they delightedly organized a Festkommers. This was a formal drinking and smoking party. It was one of the two ways of expressing esteem for a professor. The other — the ultimate honor — was a torchlight procession, which was tendered to only a few professors, and then only at the end of a long and distinguished career.

The highlight of the celebration was a speech by Klein in which he developed a magnificent comprehensive picture of Hilbert's research and teaching and its influence on the future of mathematics. "Please give me that in writing," Hilbert was heard to say afterwards.

Minkowski, back in Zürich, wrote happily:

"I have the most beautiful hopes for my future life and work!"

XII

Second Youth

After Minkowski's arrival in Göttingen in the fall of 1902, Hilbert was no longer lonely: "A telephone call, or a few steps down the street, a pebble tossed up against the little corner window of his study, and there he was, always ready for any mathematical or non-mathematical undertaking."

Instead of conducting seminars with Klein, Hilbert now conducted them with Minkowski.

On Sunday mornings the two friends regularly set out with their wives on a picnic excursion.

The Hilberts had by this time left the Reformed Protestant Church in which they had been baptized and married. It was told in Göttingen that when Franz had started to school he could not answer the question, "What religion are you?" "If you do not know what you are," he was informed by the son of the philosopher Edmund Husserl, a Jew recently converted to Christianity, "then you are certainly a Jew."

The Sunday excursions were later enlarged to include the children of both families. Most frequently the destination was a resort called Mariaspring, where there was dancing outside under the trees. Here Hilbert would seek out his current "flame" – the pretty young wife of some colleague – and whirl her around the dance floor, much to the embarrassment of the little Minkowski girls, who found his energetic dancing very old fashioned. "It was a sport to him!" They were further embarrassed when, after the music stopped, he enveloped his partner in his great loden cape and made a show of hugging and kissing her.

The frequent parties at the Hilbert house were now more enjoyable for Hilbert because of the quiet presence of Minkowski. There was dancing at these affairs too – the rug rolled away, music furnished by the gramophone a manufacturer had presented to the famous mathematics professor, the commands given in French by Hilbert. The table was always ladened with a variety of food, but the staple was talk. A subject would come up, somebody would ask Hilbert what he thought about it. Astrology, for instance. What did he think about that? Without an instant's hesitation he would answer firmly in the still uncorrupted East Prussian accent which made everything he said sound amusing and memorable: "When you collect the 10 wisest men of the world and ask them to find the most stupid thing in existence, they will not be able to find anything stupider than astrology!" Perhaps the guests would be discussing Galileo's trial and someone would blame Galileo for failing to stand up for his convictions. "But he was not an idiot," Hilbert would object. "Only an idiot could believe that scientific truth needs martyrdom — that may be necessary in religion, but scientific results prove themselves in time." Minkowski did not offer his opinion so frequently as Hilbert. When he did speak, his observation — often in the form of an appropriate quotation from *Faust* — went to the heart of the matter, and Hilbert listened. But Hilbert was always the more intrepid in expressing opinions. What technological achievement would be the most important? "To catch a fly on the moon." Why? "Because the auxiliary technical problems which would have to be solved for such a result to be achieved imply the solution of almost all the material difficulties of mankind." What mathematical problem was the most important? "The problem of the zeros of the zeta function, not only in mathematics, but absolutely most important!"

Sometimes the little Hilbert boy Franz, who was not shown at parties, would stop at the door and listen.

In front of a group, Minkowski suffered from "Lampenfieber" — in English, *stagefright*. He was still embarrassed by attention, even of much younger people; and in Zürich his shy, stammering delivery had completely put off a student named Albert Einstein. But in Göttingen ("the shrine of pure thought," as it was called) the students recognized immediately that in Minkowski they had the privilege of hearing "a true mathematical poet." It seemed to them that every sentence he spoke came into being as he spoke it.

This was, at least once, quite literally true. Lecturing on topology, Minkowski brought up the Four Color Theorem — a famous unsolved problem in that field of mathematics. (The theorem states that four colors are always sufficient to color any map in such a way that no two adjoining regions will have the same color.)

"This theorem has not yet been proved, but that is because only mathematicians of the third rank have occupied themselves with it," Minkowski announced to the class in a rare burst of arrogance. "I believe I can prove it."

He began to work out his demonstration on the spot. By the end of the hour he had not finished. The project was carried over to the next meeting of the class. Several weeks passed in this way. Finally, one rainy morning, Minkowski entered the lecture hall, followed by a crash of thunder. At the rostrum, he turned toward the class, a deeply serious expression on his gentle round face.

"Heaven is angered by my arrogance," he announced. "My proof of the Four Color Theorem is also defective."

He then took up the lecture on topology at the point where he had dropped it several weeks before. (The Four Color Theorem is still, at the date of this writing, unproved.)

Hilbert was now beginning to devote himself to integral equations with the same exclusiveness which he had earlier lavished on invariants and number fields. He had begun his investigations in a way reminiscent of his approach in the earlier subjects. In his first paper, sent as a communication to the Göttingen Scientific Society, he had presented a simple and original derivation of the Fredholm theory which exposed the fundamental idea more clearly than Fredholm's own work. There were also already glimpses of fresh, fruitful ideas to come. With an intuitive grasp of the underlying relationships existing among the different parts of mathematics and between mathematics and physics, Hilbert had recognized that Fredholm equations could open up a whole series of previously inaccessible questions in analysis and mathematical physics. It was now his goal to encompass within a uniform theoretical arrangement of equations the greatest possible domain of the linear problems of analysis.

Minkowski was occupying himself again with his beloved theory of numbers. It concerned him, according to Hilbert, that many mathematicians hardly get a breath of what he called the "special air" of number theory; and during the winter of 1903–04 he delivered a series of relatively untechnical lectures, later published as a book, in which he presented the methods he had created and some of his own most significant results in a way in which they could be easily grasped. Hilbert was just as interested as Minkowski in emphasizing "the insinuating melodies of this powerful music" – this metaphor was also Minkowski's – and when Legh Reid, one of his former American students, wrote a book on the subject, Hilbert endorsed it with enthusiasm. Number theory was "the pattern for the other sciences, . . . the inexhaustible source of all mathematical knowledge, prodigal of incitement to investigations in all other domains . . . ." A number theory problem was never dated, it was "as timeless as a true work of art." Thanks to Minkowski, Germany had recently become, once again, the number theory center of the world. "But every devotee of the theory of numbers will desire that it shall be equally a possession of all nations and be cultivated and spread abroad, especially among the younger generation, to whom the future belongs."

During 1903, Hermann Weyl arrived in Göttingen. He was an 18 year old country boy, seemingly inarticulate, but with lively eyes and a great deal of confidence in his own abilities. He had chosen the University because the director of his gymnasium was the cousin of one of the mathematics professors "by the name of David Hilbert."

"In the fullness of my innocence and ignorance," Weyl wrote many years later from the Institute for Advanced Study in Princeton, New Jersey, "I made bold to take the course Hilbert had announced for that term, on the notion of number and the quadrature of the circle. Most of it went straight over my head. But the doors of a new world swung open for me, and I had not sat long at Hilbert's feet before the resolution formed itself in my young heart that I must by all means read and study whatever this man had written."

Hilbert's "optimism, his spiritual passion, his unshakable faith in the supreme value of science, and his firm confidence in the power of reason to find simple and clear answers to simple and clear questions" were irresistible. Weyl heard "the sweet flute of the Pied Piper . . ., seducing so many rats to follow him into the deep river of mathematics." That summer he went home with a copy of the *Zahlbericht* under his arm and, although he did not have any previous knowledge of the mathematics involved, worked his way through it during the vacation.

He was a young man with a taste for words as well as for mathematics, and he found the peculiarly Hilbertian brand of thinking admirably reflected in the great *lucidity* of Hilbert's literary style:

"It is as if you are on a swift walk through a sunny open landscape; you look freely around, demarcation lines and connecting roads are pointed out to you before you must brace yourself to climb the hill; then the path goes straight up, no ambling around, no detours."

The summer months spent studying the *Zahlbericht* were, Weyl was always to say, the happiest months of his life.

It was also at this time, during Minkowski's first years, that Max Born, the son of a well-known medical researcher in Breslau, came to Göttingen on the advice of his friends, Ernst Hellinger and Otto Toeplitz. They had informed him that Göttingen was now "the mecca of German mathematics."

Born's stepmother had known Minkowski in Königsberg; and not long after his arrival, the new student was invited to lunch by the professor and presented to Guste Minkowski and the two little daughters. After lunch, Hilbert and Käthe came over and the whole party hiked to Die Plesse, a ruined castle overlooking the valley of the Leine and the red-tiled roofs of Göttingen.

Born was never to forget the afternoon.

"The conversation of the two friends was an intellectual fireworks display. Full of wit and humor and still also of deep seriousness. I myself had grown up in an atmosphere to which spirited discussion and criticism of traditional values was in no way foreign; my father's friends, most of them medical researchers like himself, loved lively free conversation; but doctors are closer to everyday life and as human beings simpler than mathematicians, whose brains work in the sphere of highest abstraction. In any case, I still had not heard such frank, independent, free-ranging criticism of all possibile proceedings of science, art, politics."

To Weyl and Born and the other students it seemed that Hilbert and Minkowski were "heroes," performing great deeds, while Klein, ruling above the clouds, was "a distant god." The older man now devoted more and more of his time and energy to the realization of his dream of Göttingen as the center of the scientific world. Before the turn of the century he had brought together economic leaders and scientific specialists in an organization called the Göttingen Association for Advancement of Applied Mathematics and Mechanics. As a result of the activities of this group (familiarly known as the Göttingen Association) the University was gradually being ringed by a series of scientific and technical institutes — the model for scientific-technological complexes which were later to grow up around various universities in America.

Sometimes Klein was now a little amusing for the seriousness with which he took himself and his many projects. It was said that he had but two jokes, one for the spring semester and one for the fall. He did not allow himself the pleasures of ordinary men. Every moment was budgeted. Even his daughter had to make an appointment to talk with her father.

Without ever making an issue of the matter, Hilbert and Minkowski saw to it that they themselves were never organized. Once, after Klein had completely filled a very large blackboard with figures on the German middle schools (the scientific education of which he was also trying to reform), he asked his colleagues if they had any comments. "Doesn't it seem to you, Herr Geheimrat," Minkowski asked softly, "that there is an unusually high proportion of primes among those figures?" On another occasion, when Klein, written agenda in hand, tried to turn the informal weekly walks of the mathematics professors into department meetings, Hilbert simply failed to turn up for the next walk. But for the most part the three men, so different in their personalities, worked together in rare harmony.

In 1904, when the Extraordinariat for applied mathematics became vacant, Klein proposed to Althoff that a full professorship in that subject be established, the first in Germany specifically designated for applied mathematics. For the new position he had in mind Carl Runge, then at Hannover. Runge was not only a distinguished experimental physicist, famous for his measurement of spectral lines, but also a first-rate mathematician, whose name is attached to the approximation of analytic functions by means of polynomials.

Runge had known and admired Klein for almost ten years and had recently become acquainted with Hilbert. "Hilbert is a charming human being," he had written his wife. "His idealism and his friendly disposition and unassuming honesty cause a person to like him very much." The possibility of associating with these two gifted mathematicians was so exciting to Runge, who had felt very much alone in Hannover, that he went to Berlin to talk to Althoff about the new position with the feeling that it was too good to be true. "However," as his daughter later wrote, "he had not reckoned with the fact that for Klein's most broad and comprehensive plans there was always more sympathy from Althoff than for the individual plan of another." The new position was his if he wanted it. The salary would be somewhat less than what he had been receiving in Hannover, however.

"But you must not let yourself be influenced by financial considerations," his wife wrote emphatically when she heard the news. "We will come through, even with a thousand Marks less, and it will not hurt either me or the children."

At the beginning of the winter semester 1904–05, Runge joined the faculty. The mathematics professors, a quartet now, took up the practice of a weekly walk every Thursday afternoon punctually at three o'clock. Klein gave up preparing agendas. The walks became pleasant informal rambles during the course of which anything, including department business, could be discussed; and, as Hilbert happily remarked, "science did not come out too short."

Runge had a gift for computation which impressed even his new colleagues. Once, when they were trying to schedule a conference several years in the future, it became necessary to know the date of Easter. Since the determination of Easter is no simple matter, involving as it does such things as the phases of the moon, the mathematicians began to search for a calendar. But Runge merely stood silent for a moment and then announced that Easter that year would fall on such and such a date.

It was equally amazing to the mathematicians how Runge was able to handle mechanics. When the Wright brothers made their first flight, he constructed a model of their plane from paper scraps which he weighted down with needles and then allowed to glide to the ground. In this way he estimated "rather correctly" the capacity of the motor, the details of which were still a secret.

When Runge arrived in Göttingen, the scientific faculty most closely connected with the mathematicians was also impressive. Physicists were Eduard Riecke and Woldemar Voigt. H. T. Simon was the head of the Institute for Applied Electricity; Ludwig Prandtl, of the Institute for Applied Mechanics; and Emil Wiechert, the Institute for Geophysics. Karl Schwarzschild was professor of astronomy.

But the stimulating society was not limited to these high-ranking men. Otto Blumenthal, who was to be distinguished for the rest of his life as "Hilbert's oldest student," was very close to the professors although he was still merely a Privatdozent. He was a gentle, fun-loving, sociable young man who spoke and read a number of languages and was interested in literature, history and theology as well as mathematics and physics. Although born a Jew, he eventually became a Christian and frequently spoke of "we Protestants."

The unusually close relationship existing between the docents and the professors is evidenced by the fact that when Blumenthal and Ernst Zermelo, another docent, wished to give some experimental lectures on elementary arithmetic, Hilbert and Minkowski regularly attended to give more authority to their project.

Zermelo was somewhat older than Blumenthal, a nervous, solitary man who preferred whisky to company. He liked to prove at this time, which was before Peary's expedition, the impossibility of reaching the North Pole. The amount of whisky needed to reach a latitude, he maintained, is proportional to the tangent of the latitude, i.e., approaches infinity at the

Pole itself. When newcomers to Göttingen asked him about his curious name, he told them, "It used to be *Walzermelodie*, but then it became necessary to discard the first syllable and the last."

It was Zermelo who had recently pointed out to Hilbert a disturbing antinomy in set theory — the same antinomy which the young English logician Bertrand Russell pointed out to Gottlob Frege as Frege was ready to send to the printer his definitive work on the foundations of arithmetic. This antinomy — a contradiction reached by using methods of reasoning which had been accepted by mathematicians and everyone else since the time of Aristotle — had to do with the commonly recognized fact that some sets are members of themselves and others are not. For instance, the set of all sets having more than three members is a member of itself because it has more than three members. On the other hand, the set of all numbers is not a member of itself, since it is not a number. But now Zermelo and Russell, independently, had brought up the question of *the set of all sets that are not members of themselves.* Since the members of this set are the sets which are not members of themselves, the set is a member of itself if, and only if, it is not a member of itself.

By 1904, after its publication by Russell, the antinomy was having — in Hilbert's opinion — a "downright catastrophic effect" in mathematics. One after another, the great gifted workers in set theory — Frege himself as well as Dedekind — had all withdrawn from the field, conceding defeat. The simplest and most important deductive methods, the most ordinary and fruitful concepts seemed to be threatened; for this antinomy and others had appeared simply as a result of employing definitions and deductive methods which had been customary in mathematics. Even Hilbert had now to admit that perhaps Kronecker had been right — the ideas and methods of classical logic were in fact not equal to the strong demands of set theory.

In the past Hilbert had believed that Kronecker's doubts about the soundness of set theory and certain parts of analysis could be removed by substituting consistency, or freedom from contradiction, for construction by means of integers as the criterion of mathematical existence and then obtaining the necessary absolute proof of the consistency of the arithmetic of real numbers. Up until the discovery of the antinomies it had been his opinion that the desired consistency proof could be achieved relatively easily by a suitable modification of known methods of reasoning in the theory of irrational numbers. But with the discovery of the antinomies in set theory, upon which much of this reasoning was based, he saw that he was going to have to alter his view. In the late summer of 1904, when the third International Congress of Mathematicians met at Heidelberg, he departed from integral equations for the moment to take up the subject of the foundations of mathematics.

It had been Kronecker's contention that the integer was the foundation of arithmetic and that construction by means of a finite number of integers was therefore the only possible criterion of mathematical existence. Hilbert still, as always, passionately opposed such a restriction of mathematics and mathematical methods. Like Cantor, he firmly believed that the essence of mathematics is its freedom; and he saw in arbitrary restriction a real danger to the science. He was sure that there was a way to eliminate the antinomies without making the sacrifices which Kronecker's views demanded. His solution, however, involved going even farther than Kronecker had.

Hilbert now insisted that the integer itself "can and must" have a foundation.

"Arithmetic is often considered to be a part of logic, and the traditional fundamental logical notions are usually presupposed when it is a question of establishing a foundation for arithmetic," he told the mathematicians gathered at Heidelberg. "If we observe attentively, however, we realize that in the traditional exposition of the laws of logic certain fundamental arithmetical notions are already used; for example, the notion of set and, to some extent, also that of number. Thus we find ourselves turning in a circle, and that is why a partly simultaneous development of the laws of logic and of arithmetic is required if the antinomies are to be avoided."

He was convinced, he told them, that in this way "a rigorous and completely satisfying foundation" could be provided for the notion of number — "number" which would include, not only Kronecker's natural numbers and their ratios (the common fractions), but also the irrational numbers, to which Kronecker had always so violently objected, but without which "the whole of analysis," in Hilbert's opinion, "would be condemned to sterility."

It was Hilbert's proposal at Heidelberg that for the first time in the history of mathematics, proof itself should be made an object of mathematical investigation.

Poincaré commented several times, unfavorably, upon the idea. The Frenchman was convinced that the principle of complete, or mathematical, induction was a characteristic of the intellect ("in Kronecker's language," as Hilbert once explained in discussing Poincaré's position, "created by God") and that, therefore, the principle could not be established except by complete induction itself.

Hilbert did not follow up the Heidelberg proposal. Instead, he continued to work on his theory of integral equations and on the side, in the company of Minkowski and at Minkowski's suggestion, began a study of classical physics.

Minkowski already had considerable technical knowledge in the field of physics; Hilbert had almost none and was familiar only with the broad outlines of the subject. Nevertheless he took it up with enthusiasm. For the second time since leaving school — the first time had been in connection with the *Zahlbericht* — he embarked on a course of "book study." Blumenthal, who had already begun what was to be a life-long study of his teacher's character and personality, was most impressed. He remembered an occasion during his own student days when in the course of his reading he had discovered to his dismay that the most beautiful development in his dissertation had already appeared in another paper. Hilbert, he recalled, had merely shrugged and said, "Why do you also know so much literature?"

Klein followed the joint physics study with interest. When he was 17, he had been assistant to Julius Plücker in Bonn. At that time he had been determined that "after the attainment of the necessary mathematical knowledge" he would devote his life to physics. Then, two years later, Plücker had died (as in Minkowski's life, when he was at Bonn, Heinrich Hertz had died). A transfer to Göttingen — where the mathematicians were a far livelier group than the physicists — had made Klein a mathematician instead of a physicist.

As the physics study progressed, Minkowski became increasingly fascinated by the riddles of electrodynamics as recently formulated in the works of H. A. Lorentz. But Hilbert did not waver in his personal concentration on integral equations. In 1904 he sent a second communication to the scientific society in which he developed a significant extension of Fredholm's idea. In his classic work, Fredholm had recognized the analogy between integral equations and the linear equations of algebra. Hilbert now went on to set up the analogue of the transformation of a quadratic form of n variables onto principal axes. Out of the resulting combination of analysis, algebra and geometry, he developed his theory of eigenfunctions and eigenvalues, a theory which, as it turned out, stood in direct relation to the physical theory of characteristic oscillations.

The spirit and significance of this work can best be glimpsed by a layman in the evaluation of it by a later student of Hilbert's:

"The importance of scientific achievement is often not alone in the new material which is added to material already on hand," Richard Courant has written. "Not less important for the progress of science can be an insight which brings order, simplicity and clarity into an existing but hard to reach area and thus facilitates or first makes possible the survey, comprehension and mastery of the science as a unified whole. We should not forget this point of view in connection with Hilbert's works in the field of analysis, . . . for [all of these] exemplify his characteristic striving to find in the solution of new problems the methods which make the old difficulties easy, to establish new connections in existing materials and to bring the many branching streams of individual investigations back into a single bed."

It was during this happy, productive time that Hilbert received yet another offer to leave Göttingen. Leo Koenigsberger would give up his own chair in Heidelberg if Hilbert would accept it.

Although Käthe favored the change, Hilbert refused.

He did not neglect, however, to use the offer to negotiate for further advantages for mathematics as the price of his remaining in Göttingen. To one of his proposals Althoff objected: "But we do not have that even in Berlin!"

"Ja," Hilbert replied happily, "but Berlin is also not Göttingen!"

XIII

The Passionate Scientific Life

At the beginning of the twentieth century mathematics students all over the world were receiving the same advice:

"Pack your suitcase and take yourself to Göttingen!"

Sometimes it seemed that the little city was entirely populated with mathematicians. But it should be mentioned that there were other people, and for some *la grande affaire* was quite different. A French journalist, choosing Göttingen as the place where he would be best able to observe German students in their natural state, was most impressed by the prison on the third floor of the Great Hall of the University. On Weender Strasse he saw, not mathematicians, but young men who "walked like lords," their caps visored in the bright colors of the duelling student fraternities, their faces usually swathed in bandages. "They leave behind them," he reported, "a nauseating odor of iodoform, which penetrates everywhere in Göttingen." But the mathematicians preferred to recount how Minkowski, walking along Weender Strasse, saw a young man pondering an obviously grave problem and patted him on the shoulder, saying, "It is sure to converge" — and the young man smiled gratefully.

The days were long past when Hilbert had delivered his lectures on analytic functions for the sole benefit of Professor Franklin. Frequently now several hundred people jammed the hall to hear him, some perching even on the window sills. He remained unaffected by the size or the importance of the audience. "If the Emperor himself had come into the hall," says Hugo Steinhaus, who came to Göttingen at this time, "Hilbert would not have changed."

Was Hilbert so because of his position as the leading mathematician in Germany? "No, Hilbert would have been the same if he had had only one piece of bread."

Born was now Hilbert's "personal" assistant. At that time it was generally customary in German universities that only professors in the experimental sciences had assistants, who helped them with the laboratory work. But Klein, shortly after he took over mathematics in Göttingen, had contrived to obtain funds for a paid clerk for the Lesezimmer. The first person to hold this position had been Arnold Sommerfeld, and the Lesezimmer clerk had quite naturally become Klein's assistant. Hilbert's assistant was still unpaid at this time.

It was a "rather vague" job, according to Born, "but precious beyond description because it enabled me to see and to talk to him every day." In the morning Born came to Hilbert's house, where he usually found Minkowski already present. Together, the three discussed the subject matter of Hilbert's coming lecture, which was often taking place that same morning.

Hilbert had no patience with mathematical lectures which filled the students with facts but did not teach them how to frame a problem and solve it. He often used to tell them that "a perfect formulation of a problem is already half its solution."

"He would take a good part of the hour to explain a question," Steinhaus recalls. "Then the formal proof, when it followed, would appear to us so natural that we wondered why we had not thought of it ourselves."

In the preparation periods with Minkowski and Born, Hilbert was interested only in the general principles which he would present to the class. He refused to prepare to the point where, as he said contemptuously, "the students could easily fill up fine notebooks." Instead, his goal was to involve them in the scientific process itself, to illuminate difficulties and "to shape a bridge to the solution of actual problems." The details of the presentation would come to him later on the rostrum.

"It was a wonderful learning time for me," Born has written of these discussions with Hilbert and Minkowski, "not only in science, but also in the things of human life. I admired them and loved them both and they did not let me feel how great was the disparity in knowledge and experience between them and me, but treated me as a young colleague."

When it was time for Hilbert to go to his lecture, Minkowski, returning home, would frequently take Born with him. It was only two blocks from Hilbert's house to the Minkowski apartment on Planck Strasse; but often, deep in conversation, they "made a long walk" before they got there. The little girls would come running to meet their father, the one who arrived first being hoisted on his back and carried into the house, clutching his thick dark hair and shrieking happily. Unlike Hilbert, whose kinship with youth did not extend to the very young, Minkowski understood and enjoyed children. He had been the one whose encouragement and play had finally got the baby Franz to talk and his letters to Hilbert had always contained some message for Franz. His own children were to remember a father who saw each of them separately for a few minutes every day so that the younger would have a chance to talk to him too. "Uncle Hilbert" they were to remember as a man "not good with children."

Because of his very general method of preparation, Hilbert's lectures could turn into fiascos. Sometimes the details did not come to him, or came wrong. He got stuck. The assistant, if he were present, might be able to step in and rescue him. "The students are confused, Herr Professor, the sign is not right." But frequently both he and the class were beyond such help. He might shrug, "Well, I should have been better prepared," and dismiss the class. More often, he was inclined to push on.

And still, it was commonly agreed in Göttingen, there was no teacher who came close to Hilbert. In his classes mathematics seemed to the students to be still "in the making"; and most of them preferred his lectures to the more perfectly prepared, encyclopedic and "finished" lectures of Klein.

Rather unexpectedly, Hilbert had quite an interest in pedagogy. He did not have a very high opinion of the abilities of the ordinary student and believed that nothing much was absorbed until it was heard several times. "Five times, Hermann, *five* times!" was the memorable advice he gave to Weyl when that young man started his lecturing career. "Keep computations to the lowest level of the multiplication table" and "Begin with the simplest examples" — these were other favorite rules. He himself tried to present important ideas in especially vivid forms, looking always for contrasts to make them more striking and memorable.

The lectures on ordinary differential equations opened with two equations on the blackboard: $y'' = 0$ and $y'' + y = 0$. "Meine Herren," he would say, "from those you can learn the entire theory, even the difference in the significance of the initial value and the boundary value problem."

"The sentence, 'All girls who are named Käthe are beautiful,' is not a universal law," he told another class. "For it is dependent upon the naming and that is arbitrary."

The difference between a purely existential statement and a specified object was illustrated by the statement, which always brought a laugh from the students: "Among those who are in this lecture hall, there is one who has the least number of hairs."

In addition to his own lectures, Hilbert regularly conducted a joint seminar with Minkowski. In 1905, after a year of physics study, they decided to devote the seminar to a subject in physics — the electrodynamics of moving bodies. Although the stimulus to the undertaking came originally from Minkowski, Hilbert took an active part and was a real partner, according to Born, "frequently clarifying and always pressing toward clarity."

For Born and the other students the seminar meetings were exciting and stimulating hours. The Fitzgerald contraction, Lorentz local-time, the Michelson-Morley interference experiment, all were thoroughly discussed and "we heard completely fantastic statements about electrodynamics."

In one of those coincidences that occur not infrequently in the history of science, the year of the seminar saw similar ideas developed in papers on electrodynamics and special relativity by a patent clerk in Bern. "But of that," Born says, "nothing was as yet known in Göttingen and in the Hilbert-Minkowski seminar the name of Einstein was never mentioned."

Born, much impressed by the ideas discussed in the seminar, considered taking a topic in this field for his dissertation. But he had fallen into disgrace with Klein in another seminar, and it was axiomatic in Göttingen that those who were out of favor with the great Felix did not fare well. So, rather than risk being examined by Klein in geometry, Born switched to astronomy. He still would have to be examined by a mathematician, but now it would be Hilbert.

Before the examination, Born asked Hilbert's advice on how he should prepare for the mathematical questions.

"In what area do you feel yourself most poorly prepared?" Hilbert asked.

"Ideal theory."

Hilbert said nothing more, and Born assumed that he would be asked no questions in this area; but when the day of the examination arrived, all of Hilbert's questions were on ideal theory.

"Ja, ja," he said afterwards, "I was just interested to find out what you know about things about which you know nothing."

After 1905, Minkowski concentrated almost entirely upon electrodynamics. The work of the Bern patent clerk became known in Göttingen, and Minkowski recalled his former student. "Ach, der Einstein," he said ruefully, "der schwänzte immer die Vorlesungen — dem hätte ich das gar nicht zugetraut." (*Oh, that Einstein, always missing lectures — I really would not have believed him capable of it!*)

Hilbert continued with his investigations in the field of integral equations. To maintain a close bond between these and his teaching, he often brought his results into lectures and seminars before they were in finished form. Often it happened then that the investigations moved forward in a kind of collaboration with his students who, as he later remarked with pleasure, "repeatedly contributed to a more precise formulation and sometimes actually to an extension of the field." In 1904, for example, he had published his theory of eigenfunctions and eigenvalues. It was still "laborious" at a crucial point. Then in 1905, in his dissertation, Erhard Schmidt laid down a new foundation for the Hilbert theory which was destined, because of its clarity and brevity, to play an important role.

It was at this time, in the year 1905, that the Hungarian Academy of Science surprised the mathematical world with the announcement of an impressive prize of 10,000 gold crowns for that mathematician whose achievement during the past 25 years had most greatly contributed to the progress of mathematics. It was to be known as the Bolyai Prize in honor of Johann Bolyai, the Hungarian who had been one of the discoverers of non-euclidean geometry, and his father Wolfgang Bolyai, a fellow student and life-long friend of Gauss.

The Academy appointed a committee of Julius König, Gustav Rados, Gaston Darboux, and Felix Klein to decide the winner; but, even before the committee met, it was clear to everyone in the mathematical world that the choice would be between two men. The final vote was unanimous. The Bolyai Prize would go to Henri Poincaré, whose mathematical career had begun in 1879 while Hilbert was still a student in the gymnasium; but the committee also voted unanimously that, as a mark of their high respect for David Hilbert, the report which they made to the Academy on their choice would treat his mathematical work to the same extent that it treated Poincaré's.

"No gold, but honor," Klein wrote Hilbert regretfully from Budapest.

Later, back in Göttingen, Klein explained to Blumenthal that the decisive factor in the award's going to Poincaré was that the Frenchman had accomplished "a full orbit of the mathematical science."

"But," Klein prophesied, "Hilbert will yet encompass as comprehensive a field as Poincaré!"

It was an auspicious time for the prophecy. Hilbert was now creating what was to be the crowning achievement of his analytic work — the theory of infinitely many variables which was to become universally known as "Hilbert Space" theory.

The generalization of the algebraic theory of quadratic forms from 2 and 3 to any finite number of variables had been popular with algebraists during the preceding century; and since a pair of variables represents a point in the plane and a triple represents a point in 3-dimensional space, they had found it convenient to move to "spaces" of higher dimension as the number of variables increased. Such generalization is, as E. T. Bell once remarked, "an almost trivial project for a competent algebraist." But with the extension to an infinite number of variables, convergence has to be considered, and the resulting analytic problem is "trivial for nobody." A still further generalization is one in which, for example, the points in the space are represented by continuous functions.

Because of its extreme generality, the problem he was now attacking seemed almost inaccessible even to Hilbert. But he approached it boldly.

"If we do not let ourselves be confused by such considerations, it goes for us as for Siegfried, before whom the fire fell back, and as reward we are beckoned on — to the beautiful prize of a methodologically uniform framework for algebra and analysis!"

In the convenient and vivid imagery of space, many analytic relations can be expressed in the terms of familiar concepts, and much that is complicated and obscure when formulated analytically becomes almost intuitively obvious when formulated geometrically. For this reason Hilbert Space theory — which for technical reasons Hilbert himself originally called "Spectral Theory" — offered an immensely suggestive language for the easy and direct expression of very abstract results. Although it yielded up many of his own results and also his methods much more easily, this was not, however, to be its chief significance.

"Decisive above all," Courant later wrote, "is the ordering and clarifying effect of such a general function theory on the entire methodology and conceptual development in analytic investigation."

At the same time that Hilbert was developing this very advanced and abstract mathematical theory, he was also teaching calculus to first year students.

His calculus class of 1906, although typical of his teaching technique at this level, was in some ways different from those which preceded it and followed it in that it took place during the great ski winter. Inspired by Runge, who had an English mother and was thus automatically the sportsman of the faculty, Hilbert and some of the younger teachers had decided to learn to ski. Equipment was imported from Norway, since none was manufactured in Germany at that time. The sessions, coached by Runge, took place on the gentle slope below Der Rohns, a popular inn.

"Well, you know, it is very nice but it is also very strenuous," Hilbert confided to Minkowski at the weekly meeting of the Mathematics Club. "This afternoon I found myself in a situation where I really did not know that I had fallen into a ditch. My two skis were up in the air and I was on my back. Then one of my skis came off and slid down the hill. So I had to take off the other ski and carry it through the deep snow. You know, that's all not so simple."

"Well," said Minkowski, who had not taken up the new sport, "why didn't you just let the second ski run down the same path as the first one? It would have landed next to it."

"Oh," said Hilbert. "Runge never thought of that!"

There was a slight incline between Hilbert's house and the Auditorienhaus, where the calculus class met; and when there was sufficient snow on the ground, Hilbert liked to ski to class. On such days he would dash breathlessly into the lecture hall, still wearing his enormous Norwegian ski boots with points on the toes and buckles on the backs, and leap to the rostrum, already talking.

It was still his custom to open a lecture by carefully reviewing what he had presented in the previous lecture. If he had spent 40 minutes on a subject the last time, he now spent 20 minutes on it. Only after that review did he take up the new subject.

"Last time we saw such and such. Now it seems that this is not applicable in the new case. How is that possible? Why does the old method break down? What is the matter? What can we do? How can we possibly get around this difficulty?"

He would go on in this vein for some time. He would also bring in ideas from other fields and mention very advanced results and recent work. The students would be fascinated by the glimpses of concepts and areas of mathematics which in the normal course of events they would not meet for several years, but at the same time they would find themselves growing increasingly impatient to get on to the subject of the day. Then, just as they were ready to give up hope, came the necessary new concept — "like a marble statue lighted up in a dark park."

"It was wonderful," says Paul Ewald, who was a student in the calculus class of 1906. "When it finally came, we felt as if we had actually seen Hilbert *create* the new concept that was needed."

By this time Hilbert's assistants were paid, and it was even sometimes possible to arrange a special "Ausarbeiter" for a large class. This particular year Hilbert had managed to wangle some non-mathematical funds for mathematics, and Ewald was hired for the job of "Ausarbeiter" and paid as "a forestry worker" for a nearby village. It was his duty to prepare a clean copy of the notes he took on Hilbert's lecture and then have it approved by the assistant, who was now Born's friend Hellinger.

Even in an elementary class like calculus it often happened that Hilbert got things garbled. Then Hellinger, looking sadly at Ewald's notes, would say, "Well, he's muddled it again and we'll just have to sit down and work it out." When the notes finally satisfied Hellinger, they were deposited in the Lesezimmer, where the students could consult them.

Ewald was to become a distinguished physicist, but he was always to say that he learned almost all the analysis he ever needed in Hilbert's calculus class and in the after-sessions with Hellinger.

In the spring semester of the calculus class, Hilbert bought a bicycle, a method of transportation which was just beginning to become popular in Göttingen, and at the age of 45 started to learn to ride.

Skiing was a temporary enthusiasm; but bicycling, like walking and gardening, became a regular accompaniment to his creative activity. He still preferred to work outdoors. Now the bicycle was always nearby. He would work for a while at the big blackboard which hung from his neighbor's wall. Then suddenly he would stop, jump on the bicycle, do a figure-eight around the two circular rose beds, or some other trick. After a few moments of riding, he would throw the bicycle to the ground and return to the blackboard. At other times he would stop what he was doing to pace up and down under his covered walk-way, his head down, his hands clasped behind him. Sometimes he would interrupt his work to prune a tree, dig a little, or pull some weeds. Visitors arrived constantly at the house and the housekeeper directed them to the garden, saying, "If you don't see the professor, look up in the trees." Usually the first word that Hilbert spoke revealed that in spite of appearances he was working toward the solution of some specific mathematical problem with the greatest ardor. He would continue to pursue his train of thought, but aloud now, unless the visitor had come with a problem of his own. In that case he would talk with interest and enthusiasm about that.

Richard Courant, who had recently joined the Breslau group of Born, Hellinger and Toeplitz, often observed the activity of Hilbert in the garden from the balcony of his nearby room. "A fantastic balance," it seemed to him, "between intense concentration and complete relaxation."

The year after Hilbert presented his theory of infinitely many variables to the scientific society, Erhard Schmidt, by this time a Privatdozent at

Berlin, published his own very simple and beautiful solution-method which, like his dissertation, again carried forward his teacher's work.

Thus, in Göttingen, the life of science proceeded, unforgettable to those who lived through it, but unnoted by visiting journalists.

XIV

Space, Time and Number

It was a halcyon time. Göttingen seemed a half forgotten kingdom of the past. The state of Hannover, of which it was a part, had been defeated and annexed by Prussia in 1866; but forty years later the remnants of the Hannoverian nobility still quietly resisted the rule of the victor. The mark of George II's adopted country remained on his German university as thoroughly as his own guttural accent had remained on his English. The houses that lined Prinzenstrasse were those of dukes and princes of Hannover, but their titles were given in English rather than in German. The scientific society of Göttingen was officially known in the English fashion as "die Königliche Gesellschaft der Wissenschaften" – *the Royal Society of Science*. The British minister of war regularly spent his summers in Göttingen. To the young people who had been drawn there from all over the world because of their love of mathematics, it seemed that they had "all time". But it did not always seem so to their elders.

By 1908 Hilbert and Minkowski had been friends for a quarter of a century. Hilbert was 46; Minkowski, 44.

For a while, during the anniversary summer, Hilbert's marvelously good health and natural optimism failed him. He became very nervous and depressed.

The breakdown did not seem to have been triggered by any specific experience. Some, like Blumenthal, thought it was the consequence of the reckless physical and mental exertions of the past few years. Others saw it as characteristic of the creative worker.

"Almost every great scientist I have known has been subject to such deep depressions," Courant says. "Klein, of course, but many others too. There are periods in the life of a productive person when he appears to himself and perhaps actually is losing his powers. This comes as a great shock."

In any case, Hilbert analyzed his illness with intelligence, then set about in a deliberate manner to get well. After several months of rest at a sanitarium in the nearby Harz mountains, he was lecturing again as usual in the fall.

In contrast to Hilbert, Minkowski was at a point of great creativity during the summer of 1908. In September he presented some of his new electrodynamical results at the annual meeting of the Society of German Scientists and Physicians, which was held in Cologne. The title he chose for this talk was "Space and Time."

"The views of space and time which I wish to lay before you," he began in his quiet, hesitant voice, "have sprung from the soil of experimental physics, and therein lies their strength. They are radical. Henceforth space by itself, and time by itself, are doomed to fade away into mere shadows, and only a kind of union of the two will preserve an independent reality."

He had often told his students in Göttingen, "Einstein's presentation of his deep theory is mathematically awkward — I can say that because he got his mathematical education in Zürich from me."

In his special theory of relativity, Einstein had shown that when mechanical events are described by the use of clocks and measuring rods, the description depends on the motion of the laboratory in which the instruments are used, and had stated the mathematical relations that connect the different descriptions of the same physical event.

Now came, in Minkowski's talk at Cologne, what has been called "the great moment of geometrization." In the period of a few moments Minkowski introduced into relativity theory his own beautifully simple mathematical idea of Space-Time by means of which the different descriptions of a phenomenon can be represented mathematically in a very simple manner.

"Three-dimensional geometry becomes a chapter in four-dimensional physics."

"Now you know why," he told his listeners, "I said at the outset that space and time are to fade away into shadows, and only a world in itself will subsist."

Among the members of the audience was Max Born, whose interest in relativity had been re-aroused by the recent works of Einstein. Minkowski wanted Born to return to Göttingen as his collaborator. He needed someone with the knowledge of optics which Born had. But first he wanted his former pupil to become more familiar with his own new ideas in the field. He sent Born back to Breslau with his latest electrodynamical work.

In Minkowski's work the young man found laid out "the whole arsenal of relativity mathematics . . . as it has been used every day since then by every theoretical physicist." Not until the beginning of December did he consider himself prepared to return to Göttingen.

"There followed several weeks when I saw Minkowski every day and talked with him. It was a happy time, full of scientific excitement, but also rich in experience of a personal sort, the beginning of a true friendship so far as the difference of age and experience permits this word to be used."

After they had finished discussing the relativity problems, they talked about number theory: "For Minkowski, as for Hilbert, number theory was the most wonderful creation of the human mind and spirit, equally a science and the greatest of arts."

It happened that at this time, just when Minkowski had deserted number theory for electrodynamics, Hilbert, recovering from his breakdown of the summer, had been ensnared by a notorious problem in the classical theory of numbers. In 1770, Edmund Waring, an otherwise undistinguished English mathematician, had asserted, apparently without any proof, that every ordinary whole number can be represented as the sum of four squares, nine cubes, nineteen fourth powers, and so on — in general, a finite number for every *n*th power. At about the same time it had been proved in connection with another theorem that every such number can in fact be represented as the sum of four squares. It did not follow, however, that Waring, having been right in the case of the squares, was also right for the other powers. It did not even follow that every number could be represented by some finite number of cubes, fourth powers, and so on. The number needed for some or all powers greater than 2 might increase indefinitely as the numbers to be represented increased in size. Since 1770, little progress had been made toward a proof of this statement of Waring's. Recently, however, mathematicians had begun to show new interest in the problem because of the possibility of successfully applying certain analytic methods to it. Hurwitz had worked in this direction; but, like all the other mathematicians who had in the past tried to prove Waring's theorem, he too had had to give up, defeated. But Hurwitz's work had aroused Hilbert's interest in the problem. For the moment he turned away from integral equations. He began where Hurwitz had left off, even using as his starting point an identity of the kind that had been set up by Hurwitz. At the end of 1908, exactly 138 years since Waring first made his conjecture, Hilbert produced a proof of Waring's theorem.

Typically, Hilbert's proof aimed at existence rather than actual construction. However, it differed from his proof of Gordan's Theorem in that while it did not actually establish the number of *n*th powers needed for the desired representation, it did provide a method by which at least in principle an estimate could be made in each case.

It was not by any means a simple proof. In fact, as the Russian number theorist Khinchin has pointed out, it was "not only ponderous in its formal aspect and based on complicated analytical theories . . . but also lacked transparency in conceptual respects." But in view of the previous intractability of the problem, it was a remarkable achievement.

"It would hardly be possible for me to exaggerate the admiration which I feel for the solution of this historic problem," G. H. Hardy wrote when, later, he and J. E. Littlewood had produced another proof of the theorem. "Within the limits which it has set for itself, it is absolutely and triumphantly successful . . . one of the landmarks in the modern theory of numbers."

Hilbert himself was tremendously pleased and proud. "He had fought with a master of the high degree of Hurwitz," Blumenthal noted, "and gained victory with weapons from Hurwitz's own armory at a point where Hurwitz saw no possibility of success." He thought happily of communicating his result to his old friend in his next letter. But before he told Hurwitz about it, he would present the proof of Waring's Theorem to Minkowski and the members of the joint seminar at the first meeting of the new year.

Minkowski had been gone from Göttingen during the Christmas holidays, but he returned on Wednesday, January 6. The next day being Thursday, the four mathematics professors made their weekly hike, promptly at three o'clock, to the Kehrhotel on the Hainberg. In spite of the wintry hills and leafless trees, it was a pleasant excursion. The cold air rang with loud cheerful voices and laughter. Minkowski recounted "with special liveliness" his latest results in his electrodynamical work. Hilbert astonished everyone with the announcement that he would present a proof of Waring's Theorem at the next meeting of the seminar.

On Friday Minkowski delivered a regular lecture. After that he conducted a doctoral examination.

Then, on Sunday afternoon, following dinner, he was suddenly stricken with a violent attack of appendicitis. That night the decision was made to undertake the difficult operation to remove the ruptured organ.

Through Monday Minkowski's condition worsened. He was conscious and quite clear about the hopelessness of his situation. On the hospital bed he studied the proof-sheets of some of his latest work and considered whether it would be possible to turn the still unfinished part of the work to good account.

Hilbert later recalled: "He spoke his regrets upon his fate, since he still could have accomplished much; but he decided that it would be good to correct the proof-sheets so that his latest electrodynamical works could be more easily read and better understood." He said that perhaps after his death the opposition to his new ideas could be more easily overcome.

"Even on the hospital bed, lying mortally afflicted, he was concerned with the fact that at the next meeting of the seminar, when I would talk on my solution of Waring's Problem, he would not be able to be present."

At noon on Tuesday, January 12, 1909, Minkowski asked to see his family and Hilbert again. Hilbert left home as soon as he received the message; but by the time he reached the hospital, Minkowski was dead. Not having yet attained his forty-fifth year, he had been taken "in the full possession of his vital energy, in the middle of his most joyful work, at the height of his scientific creativity."

Later that afternoon Hilbert wrote a letter to Hurwitz. "My Dear Old Friend," he began. "Now you alone are that to me . . . ."

The handwriting, larger than usual, became blurred as the short note progressed. He had planned, he said, to write about "a good idea" he had obtained for the solution of Waring's Problem "from your beautiful work" in the Göttingen *Nachrichten*, "but instead you receive this sad letter." He signed himself, "Your Old Friend." Then, as if to convince them both of what had happened, he repeated in a postscript the activities of Minkowski during the week, the return from the Berlin trip on Wednesday, the happy hike to the Kehrhotel on Thursday afternoon, the lecture on Friday and the doctoral examination, the attack on Sunday, the operation on Sunday night.

"The doctors themselves stood around his bed with tears in their eyes."

On Wednesday morning the announcement was made to the students.

"I was in class when Hilbert told us about Minkowski's death, and Hilbert wept," a former student recalls. "Because of the great position of a professor in those days and the distance between him and the students it was almost more of a shock for us to see Hilbert weep than to hear that Minkowski was dead."

On Thursday afternoon there was no mathematical walk. Instead, the mathematics professors provided Minkowski's body with a final escort. Again, Hilbert noted, it was exactly three o'clock.

"The strong mathematicians were like men confounded," another student wrote to his parents after the funeral. "To all appearances Klein himself found it difficult to speak calmly. Hilbert and Runge seemed disfigured, their eyes were so red with tears."

The solution of Waring's Problem – "over the proof-sheets of which his sure eye has never passed" – was published shortly afterwards and inscribed by its author:

"To the memory of Hermann Minkowski."

XV

Friends and Students

No amount of clever negotiation at the ministry in Berlin could obtain a replacement for the friendship and scientific stimulation which Hilbert had received from Minkowski. But life had to go on.

The strain under which Hilbert worked is revealed in an incident which occurred during a lecture a short time after Minkowski's death. Among those in the audience was a young man who, in spite of the fact that the professor was obviously distraught, persisted in interrupting him with questions. Finally Hilbert snapped:

"We are not here to give you information."

"But that is what you are paid for, Herr Geheimrat!"

In the shocked, embarrassed silence that followed, Hilbert, obviously shaken and angry, waited for the offender to leave the lecture hall. The young man remained stubbornly in his seat. Finally it was the professor, his face white, who turned and walked out.

"Such an incident would never have occurred," says a man who was present, "if Hilbert had been himself."

But for the most part Hilbert succeeded in accepting his loss with the same philosophical calm with which he had seen his friend accept death. There was no recurrence of the deep depression of the summer.

He took an active part in the selection of Minkowski's successor. He and Klein agreed that they were looking for a young man, someone whose achievements were still ahead of him. Hurwitz was thus ruled out. Various young mathematicians were considered. The choice finally narrowed down to Oskar Perron and Edmund Landau, and there was considerable discussion of the respective merits of the two among the members of the mathematical faculty.

"Oh, Perron is such a wonderful person," Klein said at last. "Everybody loves him. Landau is very disagreeable, very difficult to get along with. But we, being such a group as we are here, it is better that we have a man who is not easy."

This was the decisive statement. In the spring following Minkowski's death, the 32-year-old Landau came to Göttingen as professor of mathematics.

Landau's specialty was the application of analytic methods to the theory of numbers. Already, while a Privatdozent in Berlin, he had proved a very general theorem on the distribution of the prime ideals in an arbitrary algebraic number field, corresponding to the classical Prime Number Theorem. He had also done important work in function theory and had been able to extend Picard's famous theorem in such a wholly unexpected way that even he himself had not at first believed that he was correct and had delayed publishing his paper for more than a year.

His book on the distribution of primes, which is the central problem in the analytic theory of numbers, appeared the same year that he arrived in Göttingen. "In it," G. H. Hardy wrote many years later, "the analytic theory of numbers is presented for the first time, not as a collection of a few beautiful scattered theorems, but as a systematic science." It transformed the subject, "hitherto the hunting ground of a few adventurous heroes," into one of the most fruitful fields of mathematical research.

Although most German professors at that time came from the upper middle class and were comfortably well off, Landau was very rich. When people asked him how to get to his house in Göttingen, he said simply: "You will have no difficulty in finding it. It is the finest house in town."

Soon after his arrival at the University, the Landau stories began to compete in number with the Hilbert stories.

A student consulted Landau regarding the quality of a piece of amber, *Bernstein* in German. In replying, Landau gave at the same time his opinion of the relative merits of two mathematicians named Bernstein who were in Göttingen then. "Felix," he said. If he had said, "Serge," it would have meant that he considered the amber to be of superior quality. (This judgment was not so damning as it sounds. Felix Bernstein was a very good mathematician, famous for his work in actuarial theory and statistics; but Serge Bernstein was one of the greatest Russian mathematicians of the time.)

Landau had none of Minkowski's interest in geometry or mathematical physics, and he had an absolute contempt for applied mathematics.

One time Steinhaus was describing his doctoral examination to Landau. For this he had had to be examined by an astronomer. Landau appeared to be very impressed that a student of pure mathematics could successfully answer questions put to him by an applied mathematician. "What did he ask you?" he demanded. Pleased to have gained the professor's interest in his experience, Steinhaus explained that the astronomer had asked him for the differential equations for the movements of three celestial bodies.

"Ah, so he knows *that*!" Landau exclaimed. "So he knows that."

That was Landau.

Colleagues and students disliked his arrogance and feared his wit and ruthless honesty. They gave him their allegiance and respect, however, for his fantastic diligence and the unexpected impersonality of his devotion to mathematics. "Most of us are at bottom a little jealous of progress by others," Hardy once commented, "[but] Landau seemed singularly free from such unworthy emotions."

Landau was scarcely settled in Göttingen before he suggested to a young Dane named Harald Bohr, who had solved a problem mentioned in one of his works, that he should come down and study with him.

Bohr was an unusual mathematician. In 1908, he had been a member of Denmark's runner-up Olympic soccer team, and he was probably the only person in the history of mathematics to have his examination for the doctoral degree reported on the sports page. In the years to come he was to attach his name inextricably to what are known as "almost periodic functions." But he was never to be able to go by a ball without kicking it.

To Bohr a spirit of genuine international brotherhood seemed to reign among the young mathematicians of Göttingen when he arrived there, a few months after Minkowski's death. The transfer of foreign currency was as easy as possible. No one was ever asked for a passport. The German students, especially the slightly older ones, looked after the young foreign students "with touching care."

"The grand old man . . . was Felix Klein. His imposing and powerful figure filled all, young and old, with great respect — one could almost say with awe . . . . But over the whole life in Göttingen shone the brilliant genius of David Hilbert, as if binding us all together . . . . Almost every word he said, about problems in our science and about things in general, seemed to us strangely fresh and enriching."

That spring Hilbert sent no communication on integral equations to the Göttingen Scientific Society. He and Käthe spent a great deal of time with Guste Minkowski and the little girls. He took over the general editorship of Minkowski's papers, and he began to plan a memorial talk. In preparation for this, he read over the ninety-odd letters that he had received from Minkowski since their university days. "It is like living our whole life over," he wrote to Hurwitz, "and I see how important a part you played in it."

Almost coincidentally with Minkowski's death an opportunity offered itself to bring to Göttingen distinguished visitors for the personal scientific exchange that was so necessary for Hilbert's own productivity and which in the past he had obtained from Minkowski. A Darmstadt mathematics professor named Paul Wolfskehl left in his will a prize of 100,000 Marks for the first complete proof of Fermat's Last Theorem. Until the prize was claimed, the interest accruing from this sum was to be used at the discretion of a committee of the Göttingen Scientific Society. Hilbert became the chairman of the committee, and in the April following Minkowski's death he was able to arrange that 2500 Marks be used to bring Henri Poincaré to Göttingen.

Socially and mathematically, the situation was delicate. The breakdown which had changed the entire course of Klein's career had been brought about by his competition with the young Poincaré. Now the leading mathematicians in the world were Hilbert and Poincaré, but the Bolyai Prize had gone to Poincaré. To many people in Göttingen, the Frenchman's presence was an unwelcome reminder that the mathematical world was not a sphere, with its center at Göttingen, but an ellipsoid.

Poincaré's choice of subjects for his lectures did not help the situation. He decided to speak on integral equations and relativity theory, both areas in which he had made substantial contributions, and he probably chose these topics because he knew that the Göttingen mathematicians were interested in them. But a foreign mathematician who was present was very surprised at the coolness with which the famous guest was received. "*We* were surprised," one of the Göttingen docents explained, "that Poincaré would come and talk to *us* about integral equations!"

Hilbert, however, always addressed Poincaré as "My Dear Friend" and referred to him in lectures and papers as "the most splendid mathematician of his generation." He and Käthe gave a large reception for the Frenchman and for Klein, whose sixtieth birthday fell during the visit.

Minkowski too had had the highest admiration for Poincaré; and the little Minkowski girl, seeing the great man on the steps of the house at Wilhelm Weber Strasse, dropped him a deep curtsy, as befits a little girl when she sees a king.

"What a joy it is to be a mathematician today!" Hilbert told his guests in a little speech. "Everywhere mathematics is budding, and new sprigs are blossoming. For in its applications to the natural sciences and in connection with philosophy, mathematics is becoming ever more important and is in the process of regaining its former central position!"

But in a letter to Hurwitz, thanking him for his friendly appreciation of the proof of Waring's Theorem, he wrote that it was "a bright ray in the darkness."

On the first of May he delivered his address in memory of Minkowski before a special meeting of the Göttingen Scientific Society.

With loving care he described his friend's work, recounted his successes and the appreciation which he had received from mathematicians like Hermite and Dedekind. "In spite of the fact that he was genuinely modest and willingly kept his person in the background, he nevertheless had the inner conviction that many of the works produced by him would survive those of other contemporary authors and would be admitted eventually to general appreciation. He rated the theorem which he discovered on the solvability in integers of linear inequalities, his proof of the existence of ramification numbers in number fields, and the reduction of the cubic inequality which expresses the maximum property of the sphere to a quadratic inequality equivalent to the finest achievements of the great classical mathematicians in the field of number theory combined with geometry."

He had accomplished a great deal in a short time. "Diligent he must have been!" His science had accompanied him wherever he went. "It was at all times interesting to him and in no part fatigued him, whether on an excursion or during the summer holiday, in an art gallery or a railway carriage or on the sidewalks of a big city."

Since their school days, Hilbert told his colleagues, Minkowski had been his best and most trustworthy friend:

"Our science, which we loved above everything, had brought us together. It appeared to us as a flowering garden. In this garden there were well-worn paths where one might look around at leisure and enjoy oneself without effort, especially at the side of a congenial companion. But we also liked to seek out hidden trails and discovered many an unexpected view which was pleasing to our eyes; and when the one pointed it out to the other and we admired it together, our joy was complete."

To Hilbert it seemed that his friend's nature had been like the sound of a bell, "so clear in the happiness in relation to his work and the cheerfulness of his disposition, so full in steadiness and trustworthiness, so pure in his idealistic aspirations and his life conception."

"He was for me a gift of the gods – such a one as would seldom fall to a person's lot – and I must be thankful that I so long possessed it."

In the coming months and years Hilbert tried to find congenial companionship among the advanced students and docents at Göttingen. He was quite clear about the necessity of contact with youth for his own scientific creativity.

"I seat myself with the young," he announced at one scientific meeting, "from whom there is still something to get."

One young friend of long standing was Leonard Nelson, a philosophy docent 20 years Hilbert's junior. Their acquaintance had begun several years earlier when Nelson, having received his doctor's degree at Berlin, was attempting to habilitate at Göttingen. Nelson was a young man with a strong leaning toward controversy, personal, philosophical, and political. He had incurred the dislike of Husserl, the philosophy professor; and his habilitation thesis was rejected by the majority of the Philosophical Faculty, which also included the mathematicians. Afterwards, when Nelson had received the bad news and was sitting dejectedly in his room, there was a knock on his door. "And to my astonishment," he wrote to his parents, "there was Hilbert in his own person. He invited me to supper at his house . . . ." In the next letter he reported, "Hilbert is 'racking his brains' how we can get my habilitation paper accepted." As it turned out, this project had taken even Hilbert several years; but now Nelson was a docent and he and the professor were frequently seen "walking the wall" together, deep in discussion of that area of knowledge where philosophy, mathematics and logic meet.

Another young friend, also not a mathematician, was Theodore von Kármán, who was an assistant in Prandtl's applied mechanics institute. Von Kármán was working on a project called the Zeppelin which the government wanted to test under a variety of atmospheric conditions. Many years later, when he was an influential figure in aviation and space research in the United States, he called Hilbert "the greatest mathematician in the history of science . . . for he developed the theory of integral equations into a tool which enabled scientists to make break-throughs in regions once muddy with confusion."

After Minkowski's death Hilbert revived the custom of taking a group of the young people for a long walk following the weekly meetings of the Mathematics Club.

"He was not young . . . but he had retained all his full strength and youthfulness," it seemed to the 22-year-old Bohr, "[and] his profound originality, his total lack of prejudice, one might even say of convention, made each of these meetings with him a real experience."

Several of the gifted students of the past few years were now working their way up the academic ladder.

At Hilbert's suggestion, Max Born was entrusted by Mrs. Minkowski with the editing of her husband's physics papers. One of these Born had to reconstruct from the barest notes. He also carried on his teacher's work with a paper of his own in which he presented a new and rigorous method for calculating the electromagnetic self-energy of the electron. A talk presented on this paper so impressed Voigt that he offered Born a position as a Privatdozent in the Institute of Theoretical Physics.

Hermann Weyl also became a Privatdozent about this time. Although he had already shown his mathematical abilities, he was still too shy for the socially "in-group" of mathematics. Thus it was something of a surprise to everybody when he shortly won the hand of a much sought-after young lady whose charms were such that when her father threatened to withdraw her from the University a petition begging him to reconsider was signed even by professors.

It was also at this time that the friendship between Hilbert and Richard Courant began.

Already it was clear that here was a young man who would go far, and not only in mathematics. From the age of fourteen he had lived alone, supported himself by tutoring the students of a girls' school, and ultimately accomplished the near-impossible by getting himself admitted to a university without ever having obtained a diploma from a gymnasium. Unlike most university students of that time, he was completely dependent upon himself for his support.

One day, after attending a lecture by Hilbert, Courant was surprised to be invited to tea by the professor. When he arrived, he learned that the Hilberts had a request to make of him. Franz Hilbert, who was now in his teens, was not doing well in the school in Göttingen.

("My son gets his mathematics from his mother," Hilbert sometimes said. "All else is from me.")

Mrs. Hilbert thought that perhaps Franz would do better in another school. To make sure that he would be accepted, would young Courant be willing to tutor him?

"So it happened that I spent quite a bit of time with Franz Hilbert. He was not an unintelligent or untalented boy. He was accessible. He learned a little bit and he was accepted by the new school, a very well known country school. But I was always impressed with the fact that here was a boy whose mind was like a photographic plate that you put in the developer and something very nice comes out and then, after a short while, there's a veil; it becomes increasingly cloudy and finally there is nothing left on the plate."

Already "the little Courant," as he was affectionately known, had a deep feeling for the broad scientific tradition of Göttingen. He also had a flair for the dramatic. When he received his degree in February 1910, he was not content with giving the customary kiss to the little goose girl in the fountain in the Rathaus Square. Instead, two friends hired a Droschke and circled the city, trumpeting the announcement to the citizens that Richard Courant was now Doctor of Philosophy *summa cum laude*!

During the year 1910 Courant was Hilbert's assistant.

That same year, for the first time since 1906, Hilbert sent a communication on integral equations to the Göttingen Scientific Society, his sixth and last.

"One can really say that it was through Hilbert's investigations that the true significance of integral equation theory was first exposed," Courant later wrote. "Their various relations to the most different fields of mathematics, the many-sided applicability and the inner harmony and simplicity of their structure, their unifying power in relation to numerous previously isolated investigations first became truly evident in Hilbert's work."

Since Fredholm, mathematicians from all over the world, but especially from Germany and the United States, had taken up the subject of integral equations.

But the day had quite definitely gone to Hilbert.

In Göttingen, life went on.

XVI

Physics

In the fall of 1910, the Hungarian Academy of Science announced the award of its second Bolyai Prize "to David Hilbert, who by the profundity of his thought, the originality of his methods, and the rigorous logic of his demonstrations has already exercised considerable influence on the progress of the mathematical sciences."

It fell to Poincaré, as secretary of the prize committee, to prepare the general summary of Hilbert's work to be presented to the Academy and then to be published.

Qualities he chose to single out for special mention were the variety of the investigations, the importance of the problems attacked, the elegance and the simplicity of the methods, the clarity of the exposition, and the care for absolute rigor. He appreciated the readability of the work. He also noted that the influence Hilbert had had on the progress of mathematics came, not only from his personal investigations, but also from his teaching, "by the assistance which he gives to his students and which permits them in turn, utilizing the methods created by their teacher, to contribute to our understanding."

He described in detail Hilbert's achievements (devoting the most space to the work on the foundations of geometry) and tried to place them in relation to the achievements of other mathematicians.

On the proof of Gordan's Theorem — "One cannot better measure the progress accomplished by M. Hilbert than by comparing the volume which Gordan has devoted to his demonstration to the lines with which M. Hilbert has been able to content himself."

On the new proof of the transcendence of e and π — "The ability to simplify what at first seems complex presents itself as one of the characteristics of M. Hilbert's talent."

On the work in algebraic number fields — "The introduction of ideals by Kummer and Dedekind was a considerable progress: it generalized and at the same time clarified the classical results of Gauss on quadratic forms and their composition. The papers of M. Hilbert . . . constitute a new step forward which is no less important than the former."

On the investigations on the foundations of geometry — "There are in the history of the philosophy of geometry, three principal epochs: the first is that in which the thinkers, at the head of whom we must cite J. Bolyai, developed non-euclidean geometry; the second is that in which Helmholtz and Lie revealed the role of the idea of motion and the group in geometry; the third has been opened up by M. Hilbert."

On the salvaging of the Dirichlet Principle — "It is needless to lay stress on the importance of the discoveries which extend beyond the special problem of Dirichlet [and] we should not be surprised at the number of investigators who are now engaged on the path opened up by M. Hilbert."

On the proof of Waring's Theorem — "We do not doubt that these considerations . . . when they are fully understood, will be applied to problems that extend far beyond that of Waring."

On the recent work in the theory of integral equations — "This discovery of M. Fredholm is certainly one of the most remarkable of recent times, . . . M. Hilbert had made important improvements, . . . of which one has to admire the simplicity, the sureness, and the generality."

Poincaré's report for the Bolyai Prize appeared in *Acta Mathematica* in 1911. Unsuspected by anyone at the time, he had summed up in it what was to be the totality of Hilbert's contribution to constructive mathematics. The following year Hilbert — now fifty years old — became, as far his fellow mathematicians were concerned, a physicist.

The recent work on integral equations (published as a book in 1912) had brought him to the boundary of mathematics and physics. In it he had combined many theories under one comprehensive viewpoint. The result had been much greater abstraction, unification, clarity and rigor than had existed in the past. Practically speaking, however, physicists had gained little, since in the majority of cases the old methods utilizing differential equations still remained most usable. But in the introduction to his book on integral equations Hilbert had expressed his delight in finding a field of physics where choice was not possible, the physical concepts definitely led to an integral equation as the only valid expression of the data. This was kinetic gas theory; and the presentation of a paper on the foundations of this theory in the spring of 1912 announced that Hilbert, the mathematician, had now turned his attention to physics.

In retrospect, it seemed to him that the contemporary era in physics had opened during his docent days when Hertz had established the existence of the electromagnetic waves predicted by Maxwell. Then had followed in rapid succession the discovery of x-rays by Roentgen – radioactivity by the Curies – electrons by J. J. Thomson. Max Planck had put forth the quantum theory. Einstein had enunciated the special theory of relativity. In years, there had been as many great discoveries as there had been in centuries. "And not a single one," Hilbert exulted, "has to yield in magnificence to the achievements of the past!"

But as a mathematician he was disturbed by a certain lack of order in the triumphs of the physicists. In this he was not alone. Walther Lietzmann, one of his students of an earlier day, has recalled "what discomfort we mathematicians felt in the lectures on theoretical physics when sometimes this, sometimes that principle, without proof, was placed before us and all sorts of propositions and conclusions derived from it. We perceived the pressing necessity for investigation to determine whether these diverse principles were compatible with one another and in what relation they stood."

Questions similar to these had been investigated by Hilbert in connection with his work on the axioms of geometry – the questions of the sufficiency, independence and consistency of the axioms. Now it seemed to him that the time had arrived for the project which he had proposed at Paris as the sixth problem for the twentieth century – the axiomatization of physics and the other sciences closely allied to mathematics. A few fundamental physical phenomena should be set up as the axioms from which all observable data could then be derived by rigorous mathematical deduction as smoothly and as satisfyingly as the theorems of Euclid had been derived from his axioms. But this project required a mathematician.

"Physics," Hilbert announced, "is much too hard for physicists."

It seemed a rather arrogant remark, but the physicists knew what he meant.

"Although he was only joking," one Nobel Prize winner later said, "he expressed thus something completely genuine: the respect for the difficulty of the problems which are posed in this field of pure thought, recognized only by one who has actually put all his intellectual power to overcoming such problems."

At Paris, Hilbert had specifically mentioned that in his opinion the investigation of the axioms of the theory of probability should be accompanied by a rigorous and satisfactory development of the method of mean values in mathematical physics, and in particular in the kinetic theory of gases. This is where he first attacked the new project.

Kinetic gas theory had developed on the principle that because the motion of molecules in a gas appears completely disordered it can be described statistically and effects related to pressure, density, temperature and such can be predicted on the basis of average motions. But the theory had not developed in a unified manner: different aspects were treated individually and without connection. By applying the axiomatic method and his theory of integral equations, Hilbert was able to achieve a beautifully simple, unified system and thus transform the theory into a usable and acceptable mathematical tool. ("It is interesting to note," von Kármán wrote many years later, "that this work of sixty years ago, when space flight was a science fiction dream, is today the basis of most of our engineering calculations on the behavior of man-made satellites.") The value of his investigations in this area lay not so much in the derivation of the physical theorems, which were already known, as in the insight which was gained into their structure, assumptions, and range of validity.

But in spite of Hilbert's belief in the power of the axiomatic method to bring order out of disorder, he recognized that he could not solve the problems of physics by sheer mathematical power alone. He would have to inform himself on current developments. One way to do this would be to read and study the reports of the new discoveries as they appeared in print. But this was not his way. Instead, he turned for help to his old friend, Arnold Sommerfeld.

Now in Munich, Sommerfeld was the center of the most fertile group of young physicists in Germany. It was customary at German universities that each physics professor had his own "institute" which then had its own faculty, docents, assistants and students. At Munich the largest, best equipped institute was that of Roentgen, the professor of experimental physics; the smallest, that of Sommerfeld. But when Sommerfeld had come to Munich, he had insisted that in addition to the library and desks customary for an institute of theoretical physics, there must also be facilities for experiments. After his arrival he had created a rare spirit of comradeship in his institute. While Roentgen's students worked independently and "even too much communication from door to door was not encouraged," Sommerfeld's students frequently joined him in the nearby Alps for skiing in the winter and mountain climbing in the summer — "going up and going down, talking physics all the time." During the week in Munich they gathered after lunch for "cake and physics" at a cafe near the University where formulas and diagrams of important discoveries were frequently noted down on marble tabletops and later wiped away by grumbling waitresses.

Hilbert now asked his old friend to find him a young man to be his special assistant for physics. Sommerfeld offered the job to his student, Paul Ewald, who had recently completed his dissertation on the passage of light through a crystal.

When Ewald returned to Göttingen in the spring of 1912, he was welcomed as "Hilbert's physics tutor." That, it seemed, was Hilbert's conception of the new position. He promptly assigned Ewald various topics in physics that he himself wanted to learn about.

"I remember that one thing he assigned me was the following. There was a long-standing controversy about the number of constants of elasticity in a crystal – it went back to the founders in the field – and Hilbert wanted me to read up on it and tell him who was right. So I went to the Lesezimmer and got out all the old volumes and found it all very interesting. I saw that both sides had good arguments. In fact, I couldn't find a flaw on either side, as indeed these great men had never managed to find themselves, nor the many others who had studied the problem. So I went back and reported that to Hilbert. A few years later, the whole problem – which had held up crystal physics for more than fifty years – was solved by Max Born."

Hilbert's scientific program at this time, according to Ewald, could be succinctly summarized as follows: "We have reformed mathematics, the next thing is to reform physics, and then we'll go on to chemistry." The chemistry of the day was "somewhat like cooking in a girls' high school." That was the way that Hilbert described it.

Now he planned to take up one physical theory at a time and bring it into an acceptable mathematical formulation. From the kinetic gas theory, he moved on to another field in which the concepts also led directly to integral equations. This was elementary radiation theory. During the next couple of years he published a series of papers in which he derived the fundamental theorems with the help of his theory of linear integral equations, laid down the axiomatic foundation for the theorems, and demonstrated the consistency of his axioms. The treatment of this one theory was in essence a model of the approach to physics as a whole which he had proposed at Paris.

Ewald recalls that he personally could not seem to get very "warm" about the radiation problem. He felt that Erich Hecke, Hilbert's mathematics assistant, was really much better than he in recognizing the nature of Hilbert's difficulty in regard to the various physics papers being discussed, perhaps because his mind was essentially mathematical like Hilbert's.

Hecke was to become one of the great mathematicians of his time, but he was always to look on the days when he was assistant to Hilbert as the high point of his career. For his efforts he received the sum of 50 Marks a month, approximately $12.50 in the American money of the time. One day Hilbert himself decided that this sum was inadequate, and he told Hecke that when he went to Berlin next he would take up the matter with the Minister of Culture. But after finishing his business with the Minister, who had the final say in almost all university matters, Hilbert realized that he had forgotten something. Without further ado he stuck his head out of the window of the Ministry and shouted down to Mrs. Hilbert, who was waiting for him in the park below, "Käthe, Käthe! What was that other matter I wanted to talk about?" "Hecke, David, Hecke!" Hilbert pulled in his head, turned back to the startled official, and proceeded to demand that Hecke's salary be doubled, as it was.

In May of 1912, Sommerfeld came to Göttingen under the sponsorship of the Wolfskehl Prize Commission to talk on some recent discoveries in physics. At this time he reported that Max von Laue and others had recently succeeded in passing x-rays through a crystal. This achievement revealed the true nature of x-rays and opened up a new way of studying matter. When Ewald heard about it, he recalled a conversation he had had with von Laue just before coming to Göttingen. He had gone to consult the older man about something in his dissertation but after a few minutes had found him strangely distracted.

"What would happen if you assumed much shorter waves to travel in the crystal?" von Laue wanted to know.

"It is all here in this formula," Ewald said. "You are welcome to discuss the formula, which I will copy out for you. But I have to get my thesis delivered within the next couple of days and also have to do some reviewing for my oral examination, and I don't have time."

Ewald had thought no more of this incident until he heard Sommerfeld's report on von Laue's discovery. Hilbert's conviction of the value of direct, personal scientific communication could have no more dramatic support! That afternoon Ewald hurried back to his room. He had all the formulas necessary to discuss von Laue's discovery already in his dissertation and he worked straight through the night.

But for the most part the semester as Hilbert's assistant was an easy, leisurely time and Ewald took advantage of the opportunity to observe Hilbert more closely than he had been able to do when he was a student in the calculus class in 1906.

One day Otto Toeplitz, now a Privatdozent, came to Hilbert with a paper which he had received from an auditor in his seminar.

"Most doctoral dissertations contain half an idea," he told the professor. "The good ones have one idea. But this paper has two good ideas!"

There was a problem, however. The author of the paper, whose name was Jakob Grommer, was not eligible to stand for the doctoral degree. He had never received the necessary leaving certificate from a gymnasium; in fact, he had never attended a gymnasium but had studied in a Talmudic school because he had intended to become a rabbi. It was customary in the part of eastern Europe from which he came that the new rabbi marry the daughter of the old rabbi; but when the rabbi's daughter saw the grotesque hands and feet of Grommer, who suffered from acromegaly, she refused to marry him and so ended his hopes of a rabbinical life. The rejected lover then turned to mathematics.

Hilbert took on Grommer's case "with a gleam in his eye," according to Ewald.

"If I can get a doctor's degree for this young man, who is a Lithuanian and a Jew and does not have a gymnasium diploma, then I really shall have done something!"

(Needless to say, Grommer did indeed eventually receive his degree as a doctor of philosophy.)

In spite of Ewald's liking and admiring Hilbert, he found him "a bit of an arrested juvenile." If the day was warm, Hilbert came to the lecture hall in a short-sleeved open shirt, an attire inconceivably inappropriate for a professor in that day. He peddled through the streets with bouquets from his garden for his "flames." He was also quite as likely to bear as his gift a basket of compost balanced on the handlebars. At a concert or restaurant, no matter how elegant, if he felt a draft, he borrowed a fur or a feather boa from one of the ladies present. To some people, like Ewald, it seemed that he did these things because he thought they were shocking to more conventional citizens. Others thought that he did them because they were reasonable and he was not bothered by the fact that they were contrary to accepted behavior. In any case, he had such natural dignity in whatever he did that no one laughed.

He still loved to dance, much preferring the Rector's annual ball to the formal banquet which that official gave each year for the professors and their wives. He liked pretty young ladies and delighted in explaining mathematical ideas to them. "But, my child," he would say, "you *must* understand that!" On one occasion he composed a little verse to his "beloved angel" expressing the hope that some of his favorites would receive invitations to the ball:

Lieber
Engel,
Mach mit Eile,
Dass Mareille,
Kar--, Ils--, und Wei--,
Diese drei
Auf jeden Fall
Kommen zum Rektorenball.

Then, writing the verse on a sheet of paper cut to resemble an angel, he deposited it anonymously at the Rector's office.

He liked to fancy himself as a dashing man of the world. He wore a panama hat to cover his baldness and announced that his idea of a good vacation was going on a trip with a colleague's wife. But all of the time, recalls George Pólya, who was a student in Göttingen during this period, "he looked *so innocent*."

Käthe Hilbert's reaction to her husband's numerous "flames" is contained in an anecdote of Hilbert's fiftieth birthday party. To honor the professor, some of his students composed what they called a Love Alphabet. For every letter there was a verse about one of Hilbert's loves. For "I" – "Wenn sich unsere Haare lichten / Lieben wir die kleinen Nichten. / Das ist menschliche Natur / Denkt an Ilschen Hilbert nur." (*When our hair gets thinner, we love the little nieces. Such is human nature, just think of Ilse Hilbert.*) But for "K" – no one was able to think of one of Hilbert's loves that began with "K." Then Käthe Hilbert said: "Now, really, you could think of me just for once!" Delighted, the young people immediately composed the following verse:

Gott sei Dank, nicht so genau
Nimmt es Käthe, seine Frau.

God be thanked that Käthe, his wife,
Takes not too seriously his life.

"Without Käthe," Ewald says, "Hilbert would have been truly lost." Courant adds, "Without her, he could not have lived the life he lived."

It was during that summer — the summer of Hilbert's fiftieth year — that Henri Poincaré died. He was 59 years old; for 33 years he had been incredibly productive in almost all branches of mathematics. The year before, however, he had asked the editor of a mathematical journal to accept an unfinished paper on a problem which he considered of the highest importance:

"At my age, I may not be able to solve it, and the results obtained, susceptible of putting researchers on a new and unexpected path, seem to me too full of promise, in spite of the deceptions they have caused me, that I should resign myself to sacrificing them."

It was a poignant reminder to his contemporaries that time was short. They found themselves now filled with a certain fear of death, the special characteristic of which was expressed by Vito Volterra, the leading Italian mathematician of the time, in an address on Poincaré's work:

"Among the various ways of conceiving man's affection for life, there is one in which that desire has a majestic aspect. It is quite different from the way one usually regards the feeling of the fear of death. There come moments when the mind of a scientist engenders new ideas. He sees their fruitfulness and utility, but he knows that they are still so vague that he must go through a long process of analysis to develop them before the public will be able to understand and appreciate them at their just value. If he believes then that death may suddenly annihilate this whole world of great thoughts, and that perhaps ages may go by before another discovers them, we can understand that a sudden desire to live must seize him, and the joy of his work must be confounded with the fear of having it stop forever."

With Poincaré gone, there was no longer any question about who was the greatest living mathematician — and he was up to his ears in physics.

After Ewald left Göttingen, Sommerfeld sent Alfred Landé to serve as Hilbert's physics assistant. Hilbert had moved in his lectures from radiation theory to the molecular theory of matter — the next semester he planned to devote to the theory of the electron. His treatment of these subjects was similar to that of kinetic gas theory and radiation theory, but never published.

By this time he had worked out what was to be for him a more efficient method of utilizing his physics assistant. At their first meeting he handed Landé a sheaf of separate reprints of various recently published papers on physics and instructed him to read them.

"All kinds of subjects, the physics of solid bodies and of spectra and fluids and heat and electricity, everything that came to him I was to read and, when I found it interesting, to report on it to him."

Each morning Landé came to the house on Wilhelm Weber Strasse and explained to Hilbert the subject matter of the papers that he had selected as interesting.

"That was really the beginning of my whole career as a scientist. Without Hilbert, I would probably never have read all those papers, certainly not have digested them. When you have to explain something to someone else, then you first must really understand it and be able to put it into words."

What was it like, teaching physics to Hilbert?

"Well, sometimes he wasn't an easy student at all, and I had to tell him things several times before he got them. He always tried to repeat what I told him, but in a more organized way, more simple and clear. Sometimes, right after our meeting, he would have a lecture scheduled on the subject we had been discussing. I remember often walking side by side with him from his house on Wilhelm Weber Strasse to the Auditorienhaus, still explaining things to him in the last minutes. Then in the lecture he would try to present what I had said but in his own way — his way as a mathematician, which is something quite different from that of a physicist."

In his spare time Landé studied Hilbert's book on integral equations — "a wonderful book." In the evenings he went to parties and danced with professors' daughters. His social position, he found, was much improved by his being Hilbert's physics assistant. There was only one unpleasant aspect to the job. At the Hilbert parties it was his duty as assistant to select and change the phonograph records. This was a chore which he still, after fifty years, recalls with distaste. Hilbert, who continued to receive the latest model phonograph as a gift from the manufacturer, had few classical records at this time, preferring the latest music hall "Schlagers." It was hard for Landé to find a record which he himself cared to listen to. To make matters worse, Hilbert liked his music loud. At this time volume was determined by the size of the needle, and Hilbert insisted on having the large needle. Once he went with great expectations to a concert by Caruso. But he was disappointed. "Caruso," he said, "sings on the small needle."

During 1913 Paul Scherrer came to Göttingen as a student. He found under the still surface "an intellectual life of unsurpassed intensity." It was the time when the Quantum Theory of Light was at last being taken seriously "although it could by no effort be straightened out with wave theory." This was also the year that Niels Bohr, the elder brother of Harald, put forth his planetary theory of the atom and "one tried hard to become convinced of the reality of Bohr's electron orbits in the atoms in spite of all the hesitation the physicist felt in accepting the hypothesis that the electron on its stationary orbit about the atomic nucleus does not radiate."

Niels Bohr, like his younger brother Harald, was frequently in Göttingen. People there saw Harald as *L'Allegro*; Niels, as *Il Penseroso*. But their father, a professor of medicine who was tremendously proud of both his sons, summed them up differently: "Harald is silver," he said affectionately, "but Niels — Niels is pure gold."

Hilbert enjoyed the opportunity of talking informally with Niels Bohr. Communication to others of his own discoveries and the working through in his own mind of their ideas — that was vital. Particularly now that the mathematical sciences embraced such an extensive complex of human knowledge and were in such a state of rapid and intensive progress, it seemed to him that a scientist could not be expected to acquire the information he needed through the mere reading of scientific works. The papers of the day, because of the abstractness of their thought, required in his opinion the addition of "a strong display of spirit and lively power." How valuable it would be, he thought, to bring together the leading physicists for a week of lectures and conversation!

It was long before the day of foundations and grants, but — Fermat's Last Theorem being still unproved — the interest from the bequest of the Darmstadt mathematics professor was at hand. In 1910 the money had been used to bring H. A. Lorentz to Göttingen to talk on relativity and radiation theory. In 1911 visiting lecturers had been dispensed with so that a prize of 5000 Marks could be awarded to Zermelo "for his achievements in set theory and as an aid in the full recovery of his health." In 1912 Sommerfeld had given his lectures on the recent advances in physics. Now Hilbert arranged for a week-long Wolfskehl Conference on the Kinetic Theory of Matter in the spring of 1913.

"No one who participated can forget the impression of this gathering of outstanding learned men, freely discussing the problems of their science," F. W. Levi later wrote. "Hilbert presided.... The young men who filled the hall were almost all to make their own marks later .... The prosaic lecture hall with the black iron stove at one side was the arena for an assemblage of the crown princes and the kings of science."

In the course of the Gas Week, as it was of course immediately nicknamed, Hilbert met Peter Debye, a young physics professor from Holland, who had been Sommerfeld's first assistant at Munich. Hilbert was impressed with Debye and wished that there was an appropriate place at Göttingen which could be offered to him. He proposed to the Wolfskehl Commission that the following year's interest on the prize be used to bring guest professors in the mathematical sciences to Göttingen during the summer semesters. The summer of 1914 saw the first Darmstadt Professors at Göttingen: one of these was Hilbert's former student, Alfred Haar, now a professor at Klausenburg; the other was Peter Debye.

(When people asked Hilbert why he didn't prove Fermat's Last Theorem and win the Wolfskehl Prize, he said, "Why should I kill the goose that lays the golden egg?")

That same summer it seemed also that at last Klein's plan for a separate building — *an institute* — for mathematics was about to be carried out. The land had been obtained, the funds set aside, construction scheduled to begin.

It was the summer that, in Sarajevo, the Archduke Ferdinand of Austria was assassinated by a Serbian student.

XVII

War

The long vacation began at Göttingen on the first of August. Already, Austria-Hungary had declared war on Serbia. The French army mobilized. The German army began to march through Belgium. By the end of August a dozen countries were at war.

Hilbert thought the war was stupid, and said so.

Letters came from his former students in the United States assuring him of their continued love and respect.

The enemy, recoiling before the "atrocities of the Hun" and finding itself hard put to reconcile German "barbarism" with respected German achievements in the arts and sciences, rationalized that there must be two Germanys — the military Germany of the Kaiser and the cultural Germany of Goethe, Beethoven and Kant. Germany responded with a declaration by a group of its most famous artists and scientists that they, like all the German people, were solidly behind the Kaiser. Addressed "To the Cultural World," the declaration listed the "lies and slanders of the enemy" and, beginning with the statement, "It is not true that Germany caused this war," categorically denied each one.

Those who had drawn up the declaration recognized that mathematicians, no matter how great, are not as a rule well known except to other mathematicians. The international reputations of Klein and Hilbert were such, however, that they were both asked to sign.

Klein had always been an extremely patriotic man — in 1870 he had rushed home from Paris to volunteer for the army — and now he gave permission for his name to be used without questioning the statements made in the declaration. Hilbert, on the other hand, examined the list of sentences, each one beginning, "It is not true that . . .," and, since he could not ascertain whether they were in fact true, refused to sign.

On October 15, 1914, the Declaration to the Cultural World was publicized by the German government. Those signing included such famous scientists as Ehrlich, Fischer, Nernst, Planck, Roentgen, Wasserman, Wien. One name which was conspicuously missing was that of Einstein, who was now at the Kaiser Wilhelm Institute in Berlin. According to Einstein's friend and biographer Philipp Frank, only the fact that Einstein had become a Swiss citizen saved him from being considered a traitor. Hilbert had no such protection. His refusal to sign was the more unforgivable because he was, not merely German, but Prussian. When classes resumed at the beginning of November, many people turned away from him as if he were indeed a traitor.

Most of Hilbert's mathematical colleagues were sympathetic with his action, however; even Klein shortly regretted the excess of patriotism that had led him to sign a document without first ascertaining the validity of the statements it contained. As it happened, the Declaration did not have the effect which had been hoped for. The Cultural World was shocked that respected men would put their names to such statements as "It is not true that Germany violated the neutrality of Belgium." Klein was expelled from the Paris Academy. Hilbert was allowed to remain a member.

In spite of the war, the Thursday afternoon mathematical walks continued in Göttingen. There were now more participants than there had been in Minkowski's day. Landau and Ludwig Prandtl, the professor of applied mechanics, had been added. Another participant was Carathéodory, who had returned to relieve Klein after the older man had suffered another breakdown in 1911. The easy, cultured Greek, whose family motto was "No Effort Too Much," had become a mathematician in the traditional Göttingen style. The physicist Peter Debye, who had so impressed Hilbert during the Wolfskehl Conference, had also become a regular member of the faculty by this time.

Almost all the younger men, students and docents, were gone, or soon to go. The Lesezimmer, which before the war had been crowded at all times, was almost empty. There was no such thing as an educational deferment. Brains, good grades, letters of recommendation from one's professors, great expectations carried no weight. Only a few young men had not been called. One of these was Hilbert's assistant Landé, whose bad eyesight disqualified him for the army in the beginning.

Franz Hilbert was 21 years old the year that the war began, but the army did not take him. For a long time Hilbert had held hopes for his son. There was a period when he was apprenticed to a gardener in Göttingen. "But you never can tell," Hilbert said to Ewald, who was his assistant at the time. "I was also in my youth a bit *dammelig*." Later a small job in a bookstore in Frankfurt was obtained for Franz. He did not do well at all. It became increasingly clear that he was a disturbed boy. Mrs. Hilbert worried a great deal about him and got regular reports from friends in Frankfurt.

One night, before the war, when Courant was at the house, she received a message that her son had failed to appear for work that day and no one knew where he was. Courant, who was just leaving for Berlin, volunteered to go instead with Mrs. Hilbert to Frankfurt and help her search for Franz. While they were sitting and talking to Hilbert, waiting for train time, there was a great commotion outside and Franz suddenly appeared, covered with mud and very excited. He had left the train at a village along the way and walked home. He had come to save them, he announced, from evil spirits that were after them.

"I can still see the scene before me," Courant says today. "Hilbert said to Franz, 'Oh, you stupid boy, there is nothing – there are no ghosts or devils.' Then Franz became even more upset. There was much shouting back and forth. Franz kept haranguing us about these invisible creatures that wanted to harm us. Hilbert kept hitting his hand on the table and saying, 'There are no ghosts.' It was a very weird scene. Something obviously had to be done right away. So I called the professor of psychiatry, who came and gave Franz a little injection to quiet him. Then we took him in a taxi to a clinic for mental diseases which was near the University, and he was admitted right away."

By the time they left the clinic it was morning. Courant and Hilbert went for a short walk.

"From now on," Hilbert said quietly, "I must consider myself as not having a son."

"It was very sad the way he said it, but very determined."

The tragedy of Franz Hilbert was personally disturbing to mathematicians and mathematics students at Göttingen. To explain how two such great and gifted people could have had this unhappy offspring, they began to say that the Hilberts were first cousins. This was not true, although they were cousins by marriage.

Her husband's attitude toward Franz caused Käthe Hilbert a great deal of sorrow. She, unlike Hilbert, could not consider that she no longer had a son; and the young mathematicians quickly learned that they could always get on the good side of Mrs. Hilbert with a kind word about Franz. During the war years, however, she did not let either the personal nor the general tragedy hinder her husband from functioning as a scientist. Under her skillful management, the combination of fellowship, comfort and order necessary for Hilbert to work continued to be maintained in the house on Wilhelm Weber Strasse.

Klein, in spite of poor health, also managed to maintain what Hilbert called "the mathematical arrangements." The war had ended many of the older man's activities, such as the International Commission on Schools, and had curtailed others. Several years earlier, he had brushed away the suggestion that he write the history of the mathematics of the nineteenth century: "I am too old. It needs a young man who could devote years to its preparation. No, all that I could do would be to give a few lectures on the great events; but now I am too occupied even to prepare those."

The war gave him time.

The lectures on the mathematics of the nineteenth century, delivered in the dining room of his home, were to seem to Courant, who later helped to edit them, "the perfect sweet fruit of the wisdom of Klein's old age." Courant himself never heard the lectures. He was at the front.

Hilbert, still absorbed in physics, had few students, most of them foreign. But with Debye's arrival in the summer of 1914 he had been all set to learn about the structure of matter, and he saw no reason to let the war change his plans. He asked Debye to organize a seminar on the subject. He himself opened each session with the only half-humorous request, "Now, meine Herren, you tell me, just what is an atom?"

Scherrer, working very closely with Debye at this time and himself a member of the seminar, later recalled Hilbert as "by far the most intelligent person I have ever known."

Hilbert was now primarily interested in the fundamental problems of physics and their mathematical formulation. Sometimes in the seminar he would throw out a question with the comment, "That is a purely mathematical problem." Other times he would say, "For that problem, the physicist has the great calculating machine, Nature." It was Hilbert's opinion, according to Debye, that Maxwell's equations did not attack the essence of the problem of the structure of matter — at that time the electron was the only known fundamental particle — and that equations were needed from which *it should follow* that such a particle exists.

In their daily sessions, Landé was presenting to Hilbert "in a distilled form for mathematicians" the quantum mechanics of randomly scattered events, which was at that time still in a quite primitive stage. Then in December 1914, although still not drafted, Landé decided to volunteer for the Red Cross. When Hilbert heard that his assistant was going to leave him, he was most annoyed. To Landé, his response was another example of his extreme egocentricity:

"He thought only about mathematics; and since he was considered, now that Poincaré was gone, the greatest mathematician of the time, he thought every ease was due to him, from his wife and everybody else. He squeezed me out for my physics. That was all I meant to him."

(To Landé's teacher Sommerfeld, however, Hilbert's "naive and imperative egoism" was always "egoism in the interest of his mission, never of his own person.")

Just before Christmas, Landé left Göttingen. He was in the Red Cross for two years. Then he was drafted, "because by that time they would take anybody."

In Göttingen, in the weekly Hilbert-Debye seminars, it seemed to those few students who were left that the "living pulse" of physical research was at their finger tips. The work of Einstein as he pressed forward toward a general theory of relativity was followed with great interest. Also followed was the work of others who were trying to reach the same goal. Hilbert was especially fascinated by the ideas of Gustav Mie, then in Greifswald, who was attempting to develop a theory of matter on the fundamentals of the relativity principle; and in his own investigations he was able to bring together Mie's program of pure field theory and Einstein's theory of gravitation. At the same time, while Einstein was attempting in a rather roundabout way to develop the binding laws for the 10 coefficients of the differential form which determines gravitation, Hilbert independently solved the problem in a different, more direct way.

Both men arrived at almost the same time at the goal. As the western front settled down for the winter, Einstein presented his two papers "On general relativity theory" to the Berlin Academy on November 11 and 25; Hilbert presented his first note on "The foundations of physics" to the Royal Society of Science in Göttingen on November 20, 1915.

It was a remarkable coincidence – reminiscent of Minkowski's work on special relativity and electrodynamics in the joint seminar of 1905 – but even more remarkable, according to Born (who was now in Berlin with Einstein), was the fact that it led, not to a controversy over priority, but to a series of friendly encounters and letters.

Hilbert freely admitted, and frequently stated in lectures, that the great idea was Einstein's.

"Every boy in the streets of Göttingen understands more about four-dimensional geometry than Einstein," he once remarked. "Yet, in spite of that, Einstein did the work and not the mathematicians."

On another occasion, in a public lecture, he demanded: "Do you know why Einstein said the most original and profound things about space and time in our generation? Because he had learned nothing at all about the philosophy and mathematics of time and space!"

Each man, however, was essentially a man of his own science. Originally Einstein had believed that the most primitive mathematical principles would be adequate to formulate the fundamental laws of physics. Not until much later did he see that the opposite was the case. Then it turned out that it was Minkowski, whose lectures he had found so uninteresting, who had created the mathematical conception of Space-Time which made possible his own formulation of general relativity.

"The people in Göttingen," Einstein once wryly observed, "sometimes strike me, not as if they want to help one formulate something clearly, but as if they want only to show us physicists how much brighter they are than we."

To Hilbert, the beauty of Einstein's theory lay in its great geometrical abstraction; and when the time arrived for the awarding of the third Bolyai Prize in 1915, he recommended that it go to Einstein "for the high mathematical spirit behind all his achievements."

Klein also contributed to the development of relativity theory. He was greatly impressed by Hilbert's papers on the foundations of physics. Now, almost 70, he thought he saw a way to clarify the fundamental laws of relativity theory with the old ideas of his Erlangen Program. With his knowledge of infinitesimal transformations, he was able to achieve an important abbreviation of Hilbert's calculations.

The war went on.

While the fate of Verdun hung in the balance, a young woman arrived in Göttingen. She was the daughter of the mathematician Max Noether and had been a student of his friend Gordan, the one-time "king of the invariants," now dead. She had published half a dozen papers and had lectured to her father's classes from time to time when he was ill. Now her father had retired, her mother had recently died, her brother Fritz — who had earlier been a mathematics student in Göttingen — had gone into the army. It was a time of change, and she had decided to take advantage of it.

Emmy Noether had little in common with the legendary "female mathematician" Sonya Kowalewski, who had bewitched even Weierstrass with her young charms as well as her mind. She was not even feminine in her appearance or manner. This is the first thing, even today, that the men who knew her recall. "She had a loud and disagreeable voice." "She looked like an energetic and very nearsighted washerwoman." "Her clothes were always baggy." And they still quote with delight the gentle remark of Hermann Weyl that "the graces did not preside at her cradle." But she was to be much more important to mathematics than the bewitching Sonya. Even at this time, she had an impressive knowledge of certain subjects which Hilbert and Klein needed for their work on relativity theory, and they were both determined that she must stay in Göttingen. But in spite of the fact that Göttingen had been the first university in Germany to grant a doctoral degree to a woman, it was still not an easy matter to obtain habilitation for one. The entire Philosophical Faculty, which included philosophers, philologists and historians as well as natural scientists and mathematicians, had to vote on the acceptance of the habilitation thesis. Particular opposition came from the non-mathematical members of the Faculty.

They argued formally: "How can it be allowed that a woman become a Privatdozent? Having become a Privatdozent, she can then become a professor and a member of the University Senate. Is it permitted that a woman enter the Senate?" They argued informally, "What will our soldiers think when they return to the University and find that they are expected to learn at the feet of a woman?"

Hilbert had heard what to him were similarly irrelevant arguments when he had been attempting to have Grommer's dissertation approved by the same faculty members. "If students without the gymnasium diploma will always write such dissertations as Grommer's," he had told them, "it will be necessary to make a law forbidding the taking of the examination for the diploma." Now he answered their formal argument against habilitating Emmy Noether with equal directness:

"Meine Herren, I do not see that the sex of the candidate is an argument against her admission as a Privatdozent. After all, the Senate is not a bathhouse."

When, in spite of this rejoinder, he still could not obtain her habilitation, he solved the problem of keeping her at Göttingen in his own way. Lectures would be announced under the name of Professor Hilbert, but delivered by Fräulein Noether.

The war went on.

German submarines were sinking one out of every four ships that left an English port, but the English blockade was beginning to be felt in

Germany. Food was extremely scarce. Nineteen-sixteen saw the worst famine of the war – "the Turnip Winter," as it was called. Hilbert went as often as possible to Switzerland. In the past it had seemed to him that his old friend Hurwitz had been passed over by the German universities in favor of men often "not fit to hold a candle to him." Now, in the tranquillity of Zürich, he thought that perhaps this had been for the best: the ailing Hurwitz would never have been able to stand the privations of wartime Germany.

With his native shrewdness where his own needs were concerned and with the indispensable help of his wife, Hilbert himself managed all during the war – sometimes to the amusement of his friends and colleagues – to maintain the domestic comfort which was necessary for his work.

Food was a problem. He considered meat and eggs an absolute necessity if his brains were to function at their best for mathematics. He always had great scorn for the arguments of the vegetarians: "If they had their way, we should then have to pension the oxen." His garden furnished him with fruits and vegetables. Getting meat was more difficult. But one day the Rector of the University called all the professors to the Great Hall.

"Ah, I wonder what it will be this time!" Hilbert said with anticipation to his neighbor. The last time such a meeting had been called, the University had obtained a number of geese from a peasant and had distributed these among the professors. "Perhaps now we get swine!"

The Rector began to speak. He had great news. "Our highest commander, his Majesty the Kaiser, has just declared unrestricted submarine warfare upon our enemy!"

While most of the professors clapped and cheered at this announcement, Hilbert turned back disgustedly to his neighbor: "And I thought we would get swine!" he said. "But, you see, the German people are like that. They don't want swine. They want unrestricted submarine warfare."

The lack of contact with foreign mathematicians was extremely frustrating to Hilbert. Just before the war Bertrand Russell, with A. N. Whitehead, had published his *Principia Mathematica*. Hilbert was convinced that the combination of mathematics, philosophy and logic represented by Russell should play a greater role in science. Since he could not now bring Russell himself to Göttingen, he set about improving the position of his philosopher friend Leonard Nelson.

Nelson was also a champion of the axiomatic method. His philosophical work treated two main problems: the laying of a scientific foundation for philosophy and the systematic development of philosophical ethics and a "philosophy of right." He was still firmly opposed by Husserl, the philosophy professor; and Hilbert's files contain an extremely bulky item labeled "The Nelson Affair," recording his efforts to obtain an associate professorship for Nelson during this time.

Nelson (who finally did become an associate professor, but not until after the war) later dedicated to Hilbert the three volumes of his *Lectures on the foundations of ethics* — "an attempt to open up for the sovereign domain of exact science a new province."

In the spring of 1917 the United States at last entered the war against Germany.

That same year news arrived in Göttingen that Gaston Darboux had died. Hilbert admired Darboux, not only for his mathematical work, but also for the influence he had had on mathematics in France as a man and as a teacher. He immediately prepared a memorial for publication in the *Nachrichten*. When it appeared in print, an outraged mob of students gathered in front of Hilbert's house and demanded that the memorial to the "enemy mathematician" be immediately repudiated by its author and all copies destroyed. Hilbert refused. Instead he went to the Rector of the University and threatened to resign unless he received an official apology for the behavior of the students. The apology was immediately forthcoming; the memorial to Darboux — one of the four that Hilbert wrote during his career — remained in print. (The others were to Weierstrass, Minkowski, and Hurwitz.)

At the beginning of 1918, as new men fought in the Kremlin, Germany concluded a separate peace with the Ukraine; and Alexander Ostrowski, a young Ukrainian who had been a civil prisoner in Marburg during the war, was now able to come to Göttingen. During his enforced stay in Marburg, Ostrowski had thoroughly studied Hilbert's work and also the work of Klein. Upon his arrival in Göttingen, he paid the traditional calls upon the famous mathematicians — "a right as well as a duty."

He found Klein friendly. "He spoke with me about different things and was very much astonished that I knew so much about his work." Hilbert was polite but distant. "I believe he was rather distrustful of people whom he saw for the first time."

At the beginning of the spring semester, the Central Powers, led by Germany, launched a great offensive. For a moment, to the Germans, victory seemed at hand.

Klein's friends had recently urged him to edit his collected works. At first he had refused, saying that he could not do so without the help of some younger mathematician who would know the modern point of view. After meeting Ostrowski, he felt that he had found such a person; and he started work on the project.

Klein had always had a great intuitive gift. "In his youth," Carathéodory once wrote, "he used to look at the most difficult problems and *guess* their solutions." But he had never possessed the patience to provide the logically perfect demonstration for the theorems which he was convinced were true. "He did not want to admit that the exercise of such a demonstration could be elevated into an art and that the right execution of that art is the real essence of mathematics."

This quality of Klein's made Ostrowski's work extremely difficult.

"It happened now and then that we had to discuss results in his papers which he gave without, in my opinion, sufficient proof. Then I tried to get from him a proof. I asked him, 'Well, now, how is this and this? It is a point I don't understand.' He explained it. I still did not understand. Finally I said, 'Herr Geheimrat, may I ask questions?' Now the problem was to put the questions as sharply as possible. Well, he felt himself awfully maltreated, as if somebody had pinned him up to the wall. It happened sometimes that he just stood up and went to the window for a minute to cool off. He was never disagreeable, but he really had to apply quite a lot of control."

During these months, Ostrowski also had more contact with Hilbert. He found it fascinating to study at firsthand the personality of the man whose mathematical work he had studied so thoroughly. He was particularly impressed with the way in which Hilbert had solved the problem of "how a man of exceptional quality has to arrange himself for living among people of lesser quality."

"The problem obviously offered itself to him very early . . . and he probably saw very early what the difficulties were. He was a great friend of Minkowski's, and Minkowski was a blazing star — a student in the university who wins the great prize of the Paris Academy! People must have admired Minkowski, but in a small university like Königsberg, a lot of people must have disliked him too. Minkowski was obviously a Jew and not even a Jew of German origin. At that time, I suppose, Hilbert made his first observations on the problem of a superior individual having to live with lesser beings. It is a problem that comes up quite often, and I would say that most people do not get it solved in time. They fail to recognize the existence of the problem, or else they need their superiority to overcome some complex they have. Hilbert, in my opinion, avoided the difficulties very well."

By summer the situation at the front had changed drastically. In July the German army began to retreat. News of the true state of the military affairs now reached even beyond the Rhine. The poet Richard Dehmel issued a plea to old men and young boys to stage a last resistance against the enemy. Hilbert's childhood friend, Käthe Kollwitz, now a great and also exceptionally popular artist (who had lost one of her two sons in Belgium), replied with a stirring letter to the press:

"There has been enough of dying! Let not another man fall! Against Richard Dehmel, I ask that the words of an even greater poet be remembered: *The seed for the planting must not be ground.*"

Almost exactly four years after the Declaration to the Cultural World, a new chancellor asked for an Armistice. In the early morning of November 9, 1918, the Kaiser crossed the frontier into the Netherlands.

XVIII

The Foundations of Mathematics

With a marked stiffness, a face scarred less dashingly than by a sabre, an empty sleeve or trouser leg, the young men who had been in the trenches began to return to class.

Mathematics lay before them, "fresh as May."

While they had been gone, Einstein had changed the conception of space, time and matter and had created a need for a whole new kind of geometry. In three papers totaling less than 17 pages, a young Dutchman named Brouwer had challenged the belief that the laws of classical logic have an absolute validity independent of the subject matter to which they are applied, and had proposed a drastic program to end the "foundations crisis" precipitated by the discovery of the antinomies in set theory at the beginning of the century.

The new ideas had swept up Hilbert's gifted pupil Hermann Weyl when he had returned to Zürich after service in the German army. Before the war he had become acquainted with Einstein. Now he gave a brilliant series of lectures on Einstein's ideas and published them as the book *Space, Time and Matter*, which became a scientific best seller. To his friends it seemed that Weyl "could take an intoxicated pleasure in allowing himself to be carried away or merely tossed about by the opposing currents which disturbed the period." The "foundations crisis" was irresistible to him. In 1918 he made his own contribution with his paper on the logical foundations of the continuum. He also carefully studied Intuitionism, as Brouwer's program was called.

Hilbert was disturbed by his former student's fascination with the ideas of Brouwer, who aroused in Hilbert the memory of Kronecker. At the end of the war, Brouwer was a few years older than Weyl and 20 years younger than Hilbert. He had made impressive contributions to mathematics. In 1911 he had opened a new era in topology with his proof that the dimensionality of a euclidean space is a topological invariant. His papers on point sets were considered by many to be the deepest since those of Cantor. But, like Kronecker before him, he was willing to jettison a great part of his mathematical achievement because of his philosophical ideas.

For Brouwer, neither language nor logic was a presupposition for mathematics, which, in his view, had its source in an intuition which makes its concepts and inferences immediately clear. To Weyl it was to seem that Brouwer had "opened our eyes and made us see how far generally accepted mathematics goes beyond such statements as can claim real meaning and truth founded on evidence."

Brouwer, for instance, refused to accept the Logical Principle of the Excluded Middle although, since the time of Aristotle, mathematicians had accepted without hesitation the idea that for any sentence A there are two possibilities only: either A or not-A. Now Brouwer insisted that there was a third possibility — in other words, a middle which could not be excluded.

His argument was the following:

Suppose that A is the statement "There exists a member of the set S having the property P." If the set S is finite, it is possible — in principle — to examine each member of S and determine either that there is a member of S with the property P or that every member of S lacks the property P. For finite sets, therefore, Brouwer accepted the Principle of the Excluded Middle as valid. He refused to accept it for infinite sets because if the set S is infinite, we cannot — even in principle — examine each member of the set. If, during the course of our examination, we find a member of the set with the property P, the first alternative is substantiated; but if we never find such a member, the second alternative is still not substantiated — perhaps we have just not persisted long enough!

Since mathematical theorems are often proved by establishing that the negation would involve us in a contradiction, this third possibility which Brouwer suggested would throw into question many of the mathematical statements currently accepted.

"Taking the Principle of the Excluded Middle from the mathematician," Hilbert said, "is the same as . . . prohibiting the boxer the use of his fists."

The possible loss did not seem to bother Weyl.

Brouwer's program was the coming thing, he insisted to his friends in Zürich.

"Hermann, that is mathematics in shirt-sleeves," George Pólya told him; in other words, not completely attired.

Weyl promptly offered to wager Pólya on the future of two specific propositions which would be eliminated from mathematics if Brouwer's ideas were to be accepted, as Weyl was convinced they would be — and within 20 years. The winner of the wager was to be decided by whether in 1938 Pólya was willing to admit that the two following propositions —

1. That each [non-empty] bounded set of real numbers has a precise upper bound,

2. That each infinite set of real numbers has a countable sub-set

— were in fact completely vague "and that one could ask of their truth or falsity as little as he could ask of the truth or falsity of the main ideas of Hegel's philosophy." If, by 1938, Pólya and Weyl could not agree between themselves as to the state of affairs then existing in mathematics, the determining opinion would be that of the majority of the full professors of mathematics at the Swiss Federal Institute and at the universities of Zürich, Berlin and Göttingen. The loser would then publish the conditions of the wager and the fact that he had lost it in the German Mathematical Society's *Jahresbericht.*

Hilbert himself never read a line of Brouwer's work. Increasingly, he avoided papers, preferring to get his information from lectures and conversation. Weyl was invited to Göttingen to talk to the Mathematics Club about Intuitionism.

It will be remembered that at the Heidelberg Congress, shortly after the discovery by Russell and Zermelo of a fundamental antinomy in set theory, Hilbert had sketched a mathematical-logical program which he believed would remove "once and forever" any doubts as to the soundness of the foundations of mathematics and the methods of mathematical reasoning. During the intervening years, absorbed first in integral equations and later in physics, he had apparently dropped this project. In fact, just before the war, Blumenthal, walking with the Hilberts and recalling the Heidelberg Congress, had remarked that it now seemed nothing would ever come of the idea for a "theory of proof." Hilbert had made no comment, but Mrs. Hilbert (Blumenthal was later to recall) had smiled.

Since the Heidelberg Congress there had been several important developments in the study of foundations. Zermelo had proved the well-ordering theorem and had developed his axiom system for set theory. Russell and Whitehead had published their *Principia Mathematica.* But Hilbert himself did not return, publicly at least, to the foundations of mathematics until 1917.

In the spring of that year, on a visit to Zürich, he arranged for two of the young mathematicians in the circle around Hurwitz to accompany him on a walk. One of these was Weyl's friend Pólya. The other was a reserved, shy and somewhat nervous man named Paul Bernays. To the surprise of Pólya and Bernays the subject of conversation on the walk to the top of the Zürichberg was not mathematics but philosophy. Neither of them had specialized in that field. Bernays, however, had studied some philosophy and, during his student days at Göttingen, had been close to Leonard Nelson. In fact, his first publication had been in Nelson's philosophical journal. Now, in spite of his quietness, Bernays had much more to say than the usually voluble Pólya. At the end of the walk Hilbert asked Bernays to come to Göttingen as his assistant. Bernays accepted.

That September Hilbert returned to Zürich to deliver a lecture before the Swiss Mathematical Society. It was a week or so after the third anniversary of the beginning of the war, and his first words were timely:

"As in the life of nations, the single nation can prosper only when all goes well with its neighbors and the interest of the states requires that order prevails, not only in each one of the separate states, but in the relations among the states as well — so is it also in the life of science."

The talk was devoted to a favorite subject — the importance of the role of mathematics in the sciences — and might have been entitled "In praise of the axiomatic method."

"I believe," Hilbert said firmly, "that all which is subject to scientific thought, as soon as it is ready for the development of a theory, comes into the power of the axiomatic method and thus of mathematics."

But it was also in the course of this talk that he brought up certain questions which revealed for the first time since 1904 in public utterance his continued interest in the subject of the foundations of his science:

The problem of the solvability in principle of every mathematical question.

The problem of finding a standard of simplicity for mathematical proof.

The problem of the relation of content and formalism in mathematics.

The problem of the decidability of a mathematical question by a finite procedure.

To investigate these questions — he pointed out — it would be necessary first to examine the concepts of mathematical proof.

But he himself was still not ready to enter the foundations crisis personally. There were problems at home, both personal and professional.

Franz had been released from the hospital. Little jobs were obtained for him through the connections of the University, but he was not able to keep them for long; then Mrs. Hilbert would have to bring her son home, and the peace in the house on Wilhelm Weber Strasse would be disrupted.

"Hilbert suffered very much because he couldn't work in an atmosphere of this kind of thing," Courant says. "So it was rather poisonous for him. He needed an easy, protected life. His wife of course didn't want to give up her only son, couldn't, so that was the basis of some tension between husband and wife. But Hilbert was so intelligent that there was no real danger."

The Göttingen Association had been disbanded. The Lesezimmer had huge gaps in its collection. Almost all the German publishers of scientific journals and books were pulling back. The construction of the Mathematical Institute had had to be abandonned. Klein was 70 in 1919, Hilbert approaching 60. Carathéodory had left Göttingen for Berlin, where he was again in the company of his old friend Erhard Schmidt. Peter Debye had accepted a position in Switzerland. The Mark had been steadily declining. Food was scarce, living conditions crowded. Hilbert complained to Bernays that his salary was now worth less than it had been when he was a Privatdozent in Königsberg. The future looked very bleak.

In the summer of 1919, Hilbert, vacationing in Switzerland, let it be known that he "would perhaps consider favorably," "would not be totally adverse to," and "might even be inclined to accept" a position in Bern. Under normal conditions Bern could not hope to entice Hilbert from Göttingen, but conditions were not normal. Bern saw an opportunity to add the most famous mathematician in the world to its faculty; and, by-passing the canton rule that all openings were to be advertised in the press, the University eagerly extended an offer to the great German mathematician.

It seems clear now that Hilbert never had any real intention of accepting. At the end of his career he did not even list the offer from Bern among those which he had received. He apparently wanted it only as a negotiating lever with which to improve the situation at home "for mathematics."

Basically, his personal desires were very modest. Courant recalls his saying on his fiftieth birthday, when he was at the height of his fame and influence, "From now on, I think I will indulge myself in the luxury of traveling first class on the train."

The offer from Bern seems to have had the desired effect. By August 1919 Hilbert was corresponding with the new Minister for Science, Art and Popular Education about bringing foreign guest professors to Göttingen. His earlier request for 5000 Marks for this purpose was now raised to 10,000 and, in view of the progressive inflation, "perhaps we should have at least 15,000 Marks."

It was that fall, on November 18, 1919, that Hurwitz died. Since his student days, Hilbert had admired Hurwitz and his mathematical abilities without reservation. Once, in conversation with Ostrowski, he had mentioned that in his opinion there were just two kinds of mathematicians, those who tackled and solved worthwhile problems and those who did not.

"I was surprised how definite it was for him, how there were a few people who were really good and then there were the other ones that he just did not care about. I was also surprised that he would say such a thing to me. It was almost the only time when he did not behave like a 'sage.' If he had said something like that in public, other people would begin to wonder in what group he considered them. But there was no question in which group he considered Hurwitz. At that time he mentioned to me a paper of Hurwitz's which he said had completely supplanted a paper of his own. No one else would have said that about Hilbert's work. But he said it."

For the second time Hilbert went before the Göttingen Scientific Society to deliver a memorial to a lost friend of his youth. For eight and a half years he and Hurwitz had explored "every corner" of mathematics on daily walks in Königsberg. Hurwitz, he now told his colleagues, had been "a harmoniously developed, philosophically enlightened spirit, willingly prepared to acknowledge and appreciate the achievements of others and filled with sincere joy at every scientific advance." He took comfort in the fact that, passing away in a coma, Hurwitz had not had to take leave of his family. To be spared that had been his last wish.

After Hurwitz's death there was a rumor that Hilbert had been offered Hurwitz's chair in Zürich. A group of students went to see him with a poetic petition that he remain in Göttingen: "Hilbert, gehen Sie nicht nach Zürich. / Leben da ist auch recht 'schwürich.'" (*Hilbert, don't go to Zürich, life is hard there too.*) However, no offer was forthcoming from the Swiss.

The relative weight of Hilbert's scientific interests during this period was being gauged by his assistants, Bernays for mathematics and Adolf Kratzer for physics. On the day before a lecture, both men came to Hilbert's house. As his interest moved from physics back to mathematics, so did the role played by the assistants.

"In the summer of 1920, he was concerned primarily with problems of atom mechanics," Kratzer says. "His goal here was still axiomatization. Questions were directed to me. I seemed to do most of the talking while Bernays listened. But by the winter of 1920–21 his interest had begun to change. Now his chief goal was the formalization of the foundations of mathematics on a logistic basis, and Bernays talked while I listened."

Although in his own work he was moving toward the most abstract and formal conception of mathematics, Hilbert delivered at this time a series of lectures on geometry based on an approach of strictly visual intuition. They were quite frankly designed by him to popularize mathematics with the young men returning to the University after the war.

"For it is true," he conceded, "that mathematics is not, generally speaking, a popular subject."

He saw the reason for the lack of popularity in "the common superstition that mathematics is . . . a further development of the fine art of arithmetic, of juggling with numbers . . . ." He thought he could bring about a greater enjoyment of the subject he himself enjoyed so thoroughly if he could make it possible for his listeners "to penetrate to the essence of mathematics without having to weight themselves down under a laborious course of studies." He planned instead "a leisurely walk in the big garden that is geometry so that each may pick for himself a bouquet to his liking."

The next summer Hilbert lectured on relativity theory as part of a University series for all the Faculties. Here he demonstrated, according to Born, "that only one to whom the logical structure of a difficult, complicated territory is completely clear can discourse on it successfully to a lay audience.

He enjoyed these excursions into popularization, and during the twenties frequently delivered such lectures on various subjects.

But now Hilbert was becoming increasingly alarmed by the gains that Brouwer's conception of mathematics was making among the younger mathematicians. To him, the program of the Intuitionists represented quite simply a clear and present danger to mathematics. Many of the theorems of classical mathematics could be established by their methods, but in a more complicated and lengthy way than was customary. Much — including all pure existence proofs and a great part of analysis and Cantor's theory of infinite sets — would have to be given up.

"Existential" ideas permeated Hilbert's thinking, not only in mathematics, but also in everyday life. This is illustrated by an incident which Helmut Hasse observed at this time. The Society of German Scientists and Physicians was holding its first meeting after the war in Leipzig. In the evenings at the Burgkeller there was much questioning of the type, "What about Professor K. from A., is he still alive?" The 24-year-old Hasse was seated with other young mathematicians at a table quite near to the table shared by Hilbert and his party.

"I heard him put exactly this type of question to a Hungarian mathematician about another Hungarian mathematician. The former began to answer, 'Yes, he teaches at — and concerns himself with the theory of —, he was married a few years ago, there are three children, the oldest . . . .' But after the first few words Hilbert began to interrupt, 'Yes, but . . . .' When he finally succeeded in stopping the flow of information, he continued, 'Yes, but all of that I don't want to know. I have asked only *Does he still exist*?' "

According to Brouwer, a statement that an object exists having a given property means that, and is only proved when, a method is known which in principle at least will enable such an object to be found or constructed. Thus Brouwer would not accept Hilbert's youthful proof of the existence of a finite basis of the invariant system, or many others.

Hilbert naturally disagreed.

". . . pure existence proofs have been the most important landmarks in the historical development of our science," he maintained.

The fact that Weyl was moving closer to Brouwer's position disturbed Hilbert considerably.

In 1919 Weyl had published some "long held" thoughts of his own on foundations. Then in 1920 he had delivered several lectures on Brouwer's program. In the course of one of these he had declared: "I now give up my own attempt and join Brouwer." He was never called "Brouwer's Bulldog," but he might have been. In 1921 he proceeded to use his literary gifts to put Brouwer's ideas into even wider circulation.

This was too much for Hilbert.

At a meeting in Hamburg in 1922 he came roaring back to the defense of mathematics.

The state of affairs brought about by the discovery of the antinomies in set theory was intolerable — he conceded — but "high-ranking and meritorious mathematicians, Weyl and Brouwer, seek the solution of the problem through false ways."

Weyl heard "anger and determination" in the voice of his old teacher.

"What Weyl and Brouwer do comes to the same thing as to follow in the footsteps of Kronecker! They seek to save mathematics by throwing overboard all that which is troublesome . . . . They would chop up and mangle the science. If we would follow such a reform as the one they suggest, we would run the risk of losing a great part of our most valuable treasures!"

Down the list he went of some of the treasures which would be lost if the Intuitionists' program were to be adopted:

The general concept of the irrational number.

The function. "Even the number theory function."

Cantor's transfinite numbers.

The theorem that among infinitely many whole numbers there is a smallest.

The logical principle of the excluded middle.

Hilbert refused to accept such a "mutilation" of mathematics. He thought he saw a way in which he could regain the elementary mathematical objectivity which Brouwer and Weyl demanded without giving up any of the treasures that would have to be sacrificed in their program. This was essentially the "theory of proof" which he had sketched at Heidelberg in 1904.

Characteristically, his approach was a frontal assault on the problem. As Weyl himself later had to concede, Hilbert now gave "a completely new turn to the questions of the foundations and the truth content of mathematics."

The Intuitionists had objected that "much generally accepted mathematics goes beyond such statements as can claim real meaning." Hilbert countered their argument, so Weyl claimed, by relinquishing meaning altogether.

He proposed to formalize mathematics into a system in which the objects of the system — the mathematical theorems and their proofs — were expressed in the language of symbolic logic as sentences which have a logical structure but no content. These objects of the formal system would be chosen in such a way as to represent faithfully mathematical theory as far as the totality of theorems was concerned. The consistency of the formal system — that is, mathematics — would then be established by what Hilbert called finitary methods. "Finitary" was defined as meaning that "the discussion, assertion or definition in question is kept within the boundaries of thorough-going producibility of objects and thorough-going practicality of methods and may accordingly be carried out within the domain of concrete inspection."

In this way, through methods as severely limited as any Brouwer and Weyl could demand, Hilbert believed that he could surmount the new foundations crisis and dispose of the foundations questions *once and for all.*

In the year of his sixtieth birthday he set out to save the entirety of classical mathematics by what his former student was to call "a radical reinterpretation of its meaning without reducing its inventory."

The ghost of Kronecker seemed to rise before him in the program of the Intuitionists, and the violence with which he slashed out at it was (as Weyl was quick to point out) oddly in contrast to the confidence with which he predicted its ultimate failure:

"I believe that as little as Kronecker was able to abolish the irrational numbers . . . just as little will Weyl and Brouwer today be able to succeed. Brouwer is not, as Weyl believes him to be, the Revolution – only the repetition of an attempted *Putsch*, in its day more sharply undertaken yet failing utterly, and now, with the State armed and strengthened. . ., doomed from the start!"

XIX

The New Order

Hilbert was 60 years old on January 23, 1922.

Naturwissenschaften, the German equivalent of the British scientific weekly *Nature*, dedicated its last issue in January to the birthday. The frontispiece was a photograph of Hilbert sitting in a wide-armed wicker chair. He had not changed much over the years, but time had honed the intelligence and concentration in his face until, in age, he was a more impressive looking man than he had been in his youth.

Otto Blumenthal led off the issue with a sketch of Hilbert's career and of his character. As the "oldest student," Blumenthal had observed his Doctor-Father carefully for almost a quarter of a century. Now the pattern of Hilbert's life seemed to stand out in relief. His career had developed at the hand of problems. Then, with the work on the foundations of geometry, the axiomatic method had become so much a part of him that it, like the problems, had always accompanied him and led him. To Blumenthal, now, the most striking aspect of Hilbert's life seemed to be the remarkable continuity of progress. Immediately upon solving one problem, he had applied himself to the next. Indeed, it might seem to some who did not know him that he had been all mathematician, a logical machine and problem solver, a creature of pure thought.

"But I believe that Hilbert himself would wish to be judged differently," Blumenthal wrote. "The longer I know him and the more I learn about him, I see him as a wise human being who, after he first became conscious of his power, has unfailingly kept before himself a supreme goal toward which he strives on a well-plotted course: the goal of a unified view of life, at least in the specific field of the exact sciences."

There were other articles by former pupils on the five main areas in which Hilbert had worked — algebra, geometry, analysis, mathematical physics, and the philosophy of mathematics. (An article entitled "Hilbert and Women" was prepared by Courant and his friend Ferdinand Springer but, Courant recalls, "we didn't get it finished in time.")

There was also a birthday banquet, and the 73-year-old Klein, now confined to a wheel chair, presented the honored professor with the copy of the Vortrag which young Dr. Hilbert had given in 1885 in Klein's seminar at Leipzig.

The celebration marked in a sense the passing of the old order at Göttingen. After the war Richard Courant had returned to the University as an associate professor. When Erich Hecke had left, Courant had become a professor. In the coming years he was to fill the place of Klein.

Courant was an entirely different personality from the old Jupiter. Little, gnomish-faced, soft-voiced, he was never described as "olympian." Rather, his students remember "how he could present a picture of utter helplessness and indecision, how he could grumble almost inaudibly, how he could interfere and guide by non-interference and finally obtain the unfailing attachment and devotion of all his associates."

Courant was unusually democratic for a German professor. Even his books were in part the result of group effort. What were known as "Proofreading Festivals" regularly took place at a long table with all the assistants participating. Among these at various times were Willy Feller, Kurt Friedrichs, Hans Lewy, Otto Neugebauer, Franz Rellich.

"Red ink, glue and personal temperament were available in abundance," recalls Otto Neugebauer, who held the important and influential post of "head assistant." "Courant had certainly no easy time in defending his position and reaching a generally accepted solution under the impact of simultaneously uttered and often widely diverging individual opinions about proofs, style, formulations, figures, and many other details. At the end of such a meeting he had to stuff into his briefcase galleys, or even page proofs, which can only be described as Riemann surfaces of high genus; and it needed completely unshakable faith in the correctness of the uniformization theorems to believe that these proofs would ever be mapped on *schlicht* pages."

Like Klein, however, Courant was in the broad mathematical-physical tradition of Göttingen. The real core of his work was to be (as Neugebauer later saw it) "the conscious continuation and ever-widening development of the ideas of Riemann, Klein and Hilbert and the insistence on demonstrating the fundamental unity of all mathematical disciplines."

When Courant took over from Klein, the students of mathematics and theoretical physics still attended classes in the only classroom building of the University, the three-story Auditorienhaus where Weender Strasse crossed the old city wall. The third floor of this building remained the heart of the mathematical life: the common room where the Mathematics Club held its weekly meetings — the Lesezimmer with the mathematical books and journals on open shelves as Klein had decreed — the Room of the Mathematical Models where the students gathered outside the main lecture hall. A large wooden cabinet contained the entire administrative apparatus of mathematics at Göttingen — the stamps and the stationery. This was where Courant took his first revolutionary step. He applied to the Minister of Culture for permission to change the heading on the stationery from "Universität Göttingen" to "Mathematisches Institut der Universität Göttingen." After an appropriate delay, he received permission for this change.

"They don't know how much that will cost them," the head of the new Institute said softly.

Thus, at Göttingen, the new order began.

The problem of publication, so important for the progress of science, was already being solved by Courant. During the war a personal bond had developed between the Göttingen mathematicians and the publisher Ferdinand Springer. After the war (as Hilbert later described it) "under the impact of Klein's personality and my active influence, Dr. Springer placed his energy and his firm at the disposal of mathematics." Courant and Springer became close friends; and as a result of their combined efforts, scientific publishing in Germany began to return to normal.

In addition to Courant, there was another former pupil whom Klein and Hilbert wanted to see back at Göttingen. This was Hermann Weyl. In 1922 — the same year in which Hilbert had delivered his polemic against the Intuitionists at Hamburg — an offer went to Weyl.

Like Courant, Weyl was still in his thirties. As a result of the popularity of his book on relativity theory, which had gone into five printings in five years, and of his active participation in the controversy over foundations, he was perhaps the most generally known of his generation of mathematicians. But he also had already behind him impressive solid achievements in mathematics and mathematical physics. He was now at the height of his creative powers. A great stream of papers gushed forth, not only on his main mathematical themes, but on any mathematical topic that interested him. And it was not just mathematics that interested Weyl. There was philosophy. Art. Literature. He was convinced that the problems of science could not be separated from the problems of philosophy; also convinced that mathematics — like art, music and literature — was a creative activity of mankind. He loved to write, and wrote well. It has been said that no mathematical papers of the century express as vividly their author's personality. "Expression and shape are almost more to me than knowledge itself," he said once. And another time: "My work has always tried to unite the true with the beautiful; and when I had to choose one or the other, I usually chose the beautiful."

Weyl respected and loved Klein and Hilbert. He was dedicated to the Göttingen tradition. But he did not immediately agree to return to his old university. Still debating his decision at the last possible moment, he marched his wife around and around the block of their home in Zürich. When it was nearly midnight, he decided that he would accept the offer from Göttingen. Hurrying off to send the telegram, he returned a few hours later, having wired instead his refusal.

"I could not bring myself," he explained, "to exchange the tranquillity of life in Zürich for the uncertainties of post-war Germany."

Life in Germany was indeed uncertain. A period of violent unrest had followed the surrender. Then the people had elected a national assembly, which had met at Weimar and drawn up a republican constitution; but the new government was constantly under attack. Monarchists wanted to restore the empire. Communists wanted an experiment in the Russian style. National Socialists demanded a dictatorship, the rearming of Germany, and the tearing-up of the Treaty of Versailles. "Germans will have to get used to politics just as the cavemen had to get used to soap and water," Hilbert observed.

It was during this uncertain time that Courant began to bring to physical reality Klein's old dream of a great Mathematical Institute at Göttingen.

The Mark had been steadily declining in value. By 1922 the new government was issuing paper money to meet its needs, and inflation was well under way. The price of a volume of the *Annalen*, which had been 64 Marks in 1920, had doubled by the beginning of 1922. By the end of the year it was 400 Marks. By 1923 it had gone to 800 Marks; by the end of 1923, 28,000 Marks. The fees which the students paid at the beginning of the semester for lectures were virtually valueless by the end of the semester when the University turned them over to the Privatdozents. The Wolfskehl Prize of 100,000 Marks was shortly to be worth no more than a few scraps of paper (but in 1921 the interest from the Prize could still be used to bring Niels Bohr for a series of lectures — "the Bohr Festival Week").

Courant, to emphasize his Kleinian interest in applied as well as pure mathematics, had equipped his new institute with one of the early electrically driven desk computers. Its range of 19 digits now turned out to be just about right to handle the inflationary currency. Salaries and prices were expressed by basic numbers which were then multiplied by a rapidly increasing coefficient $c(t)$ such that the result represented the value expressed in Marks at a given moment t. Salaries were then computed each week on the basis of the current value of $c(t)$, which was obtained confidentially from the government. Now Courant offered to lend his computer to the University for the privilege of being given the value $c(t)$ when it was received, hours before its publication in the newspapers. By this simple method he greatly increased the purchasing power of the funds budgeted for mathematics. For the most part he used the extra money so obtained to fill the great gaps which had developed in the Lesezimmer collection during the war. For with Courant, as with Klein, the Lesezimmer was the center around which mathematics at Göttingen revolved.

What the Lesezimmer meant to the students has been described by B. L. van der Waerden, who after completing the university course at Amsterdam had come to Göttingen on the recommendation of Brouwer. Van der Waerden was a gifted young man. His father, a high school teacher, had once taken away his mathematical books because he thought the boy should be out playing with the other boys in the fresh air. He returned the books, however, when he discovered that his son had invented trigonometry on his own and was using the names and notations he had made up instead of the traditional ones.

In Göttingen, van der Waerden spent much of his time in the Lesezimmer. There had been nothing at all like it in Holland. Today he recalls how his regular lunching and walking companion, Helmuth Kneser, "used to start on a certain subject and make a few remarks which I couldn't understand at all. Then I would say to him that I would like to learn about that subject. Where could I find out about it? So he would give me the names of some books which I could find in the Lesezimmer. A day or so later I would be able to answer his questions and also make some significant remarks of my own, and then I learned much more." Sometimes, while van der Waerden was looking for a book "by Author A," he would find next to it a book "by Author B" which was even more interesting and useful. "In this way, I learned more in weeks or months in the Lesezimmer than many students learn in years and years."

In 1923 the inflation ended abruptly through the creation of a new unit of currency called the Rentenmark. Although Hilbert remarked sceptically, "One cannot solve a problem by changing the name of the independent variable," the stability of conditions was gradually restored.

Students again began to come to Göttingen from all over the world.

The University was, thanks to Landau, the center of the great number theory activity of the 1920's — "the beginning," it has been said, "of an era of arithmetic comparable to that inaugurated by Gauss in 1801." Two problems seemed to attract the most interest. One of these was the hypothesis of Riemann concerning the zeros of the zeta function, which Hilbert had listed as the eighth of his Paris Problems. The other was the determination of the exact values for the number of *n*th powers in Waring's Theorem, work on which had been opened up by Hilbert's proof of the theorem in 1909. Waring's conjecture had turned out to be, according to mathematical historians, "one of those problems that have started epochs in mathematics."

Harald Bohr and G. H. Hardy were frequent visitors in Göttingen, usually on their way to Denmark or to England to visit one another. When Hardy would leave Bohr, to return home over the choppy North Sea channel, he always mailed him a card announcing, "I have a proof for the Riemann hypothesis!" — confident, Hardy said, that God — with whom he waged a very personal war — would not let Hardy die with such glory.

There is a Hilbert story in connection with the Riemann hypothesis which, although perhaps apocryphal, must be included. According to this story, Hilbert had a student who one day presented him with a paper purporting to prove the Riemann hypothesis. Hilbert studied the paper carefully and was really impressed by the depth of the argument; but unfortunately he found an error in it which even he could not eliminate. The following year the student died. Hilbert asked the grieving parents if he might be permitted to make a funeral oration. While the student's relatives and friends were weeping beside the grave in the rain, Hilbert came forward. He began by saying what a tragedy it was that such a gifted young man had died before he had had an opportunity to show what he could accomplish. But, he continued, in spite of the fact that this young man's proof of the Riemann hypothesis contained an error, it was still possible that some day a proof of the famous problem would be obtained along the lines which the deceased had indicated. "In fact," he continued with enthusiasm, standing there in the rain by the dead student's grave, "let us consider a function of a complex variable . . . ."

During this period there came to Göttingen a big, shy boy who was to be an outstanding number theorist — in this area of mathematics, the Minkowski of the new generation. He had refused to serve in the army and had been confined in a mental institution which was located next to the clinic owned by Landau's father. Thus young Carl Ludwig Siegel, extremely gifted but without money, had become acquainted with the Göttingen professor. To Siegel, Landau presented a quite different picture from the spoiled cherub whom Norbert Wiener saw at about this same time.

"If it had not been for Landau," Siegel says simply, "I would have died."

When, however, Siegel came to Göttingen as a student in 1919, he worked almost entirely alone. "I was very eager to show what I could do by myself." He had no direct personal contact with Hilbert, but he was always to remember a lecture on number theory which he heard from Hilbert at this time. Hilbert wanted to give his listeners examples of the characteristic problems of the theory of numbers which seem at first glance so very simple but turn out to be incredibly difficult to solve. He mentioned Riemann's hypothesis, Fermat's theorem, and the transcendence of $2^{\sqrt{2}}$ (which he had listed as his seventh problem at Paris) as examples of this type of problem. Then he went on to say that there had recently been much progress on Riemann's hypothesis and he was very hopeful that he himself would live to see it proved. Fermat's problem had been around for a long time and apparently demanded entirely new methods for its solution — perhaps the youngest members of his audience would live to see it solved. But as for establishing the transcendence of $2^{\sqrt{2}}$, no one present in the lecture hall would live to see that!

The first two problems which Hilbert mentioned are still unsolved. But less than ten years later a young Russian named Gelfond established the transcendence of $2^{\sqrt{-2}}$. Utilizing this work, Siegel himself was shortly able to establish the desired transcendence of $2^{\sqrt{2}}$.

Siegel wrote to Hilbert about the proof. He reminded him of what he had said in the lecture of 1920 and emphasized that the important work was that of Gelfond. Hilbert was frequently criticized for "acting as if everything had been done in Göttingen." Now he responded with enthusiastic delight to Siegel's letter, but he made no mention of the young Russian's contribution. He wanted only to publish Siegel's solution. Siegel refused, certain that Gelfond himself would eventually solve this problem too. Hilbert immediately lost all interest in the matter.

After a semester in Hamburg with Hecke, who was now a professor there, Siegel returned to Göttingen as assistant to Courant and later became a Privatdozent. The money he earned was so little that Courant, who wanted a cycling companion, had to arrange an extra stipend so that Siegel could afford to buy a bicycle.

Courant liked to keep Klein and Hilbert in touch with the gifted young people. It was through him that Siegel had his first personal contacts with the famous mathematicians of Göttingen. Because of the post-war housing shortage he lived for a while at Klein's house. But even living under the same roof he felt the distance which people had always felt between themselves and Klein. He worried constantly that he "would say the wrong thing." Later he was taken by Courant to swim in that part of the Leine river which was roped off for the faculty. He met Hilbert in the little shed where the professors changed into their bathing suits. Young Siegel, Courant explained to Hilbert, had recently found another proof of a theorem of Hecke's connected with the Riemann hypothesis. Hilbert was very enthusiastic. "He always liked to make young people feel hopeful." In the bathhouse there with Hilbert, Siegel felt none of the constraint he had felt in Klein's house.

Soon after this meeting with Hilbert, Siegel was asked by Courant to referee a paper for the *Annalen*, of which Hilbert was still one of the principal editors. The young man found the paper inaccurate in many places and, even where accurate, unnecessarily roundabout in its methods. He reported to Hilbert that in his opinion the paper was not publishable.

"No, no, I must publish it!" Hilbert insisted. "In 1910 this man was a member of the committee that gave me the Bolyai Prize, and now I simply cannot refuse to publish his paper! Take it and change whatever should be changed. But I must publish it!"

The paper appeared in an improved form in the *Annalen*. Several months later, when Siegel was sure that Hilbert had forgotten all about the matter, a package was delivered to his rooms. It contained the two volumes of Minkowski's collected works, inscribed "With friendly thoughts from the editor."

One of the most fertile circles of research in post-war Göttingen revolved around Emmy Noether. The desired position of Privatdozent had at last been obtained for her in 1919. This was still the lowest possible rank on the university scale, not a job but a privilege. But Emmy Noether was delighted with the appointment. In the thirteen years which had passed since she had had to defend her doctoral dissertation before Gordan, she had come a long way. Already she had achieved important results in differential invariants, which the Soviet mathematician Paul Alexandroff was to consider sufficient to secure her a reputation as a first-rate mathematician, "hardly less a contribution to mathematical science than the notable researches of Kowalewski." She herself was always to dismiss these works as standing to the side of her main scientific path, on which at last she was now, at the age of 39, taking her first step — the building up on an axiomatic basis of a completely general theory of ideals. This work would have its source in the early algebraic work of Hilbert, but in her hands the axiomatic method would become no longer "merely a method for logical clarification and deepening of the foundations [as it was with Hilbert] but a powerful weapon of concrete mathematical research." Gordan's picture still hung over her desk in Göttingen; but although she had been so thoroughly under his influence in her youth that her dissertation had concluded with a table of the complete system of covariant forms for a given ternary quartic and had contained more than three hundred forms in symbolic representation — a maiden work which she in later years dismissed as "Formelgestrüpp!" — *a jungle of formulas* — she was destined in the next decade to make Hilbert's "theology" look like mathematics.

In 1922 she became a "nicht beamteter ausserordentlicher Professor" — an unofficial extraordinary, or associate, professor. There were no obligations connected with this new title — and no salary, such an extraordinary professor being considered more than usually inferior to an ordinary professor. The title could be explained only by a Göttingen saying to the effect that "an extraordinary professor knows nothing ordinary and an ordinary professor knows nothing extraordinary." By this time, however, inflation had so reduced the students' ability to pay fees that if the Privatdozents were not to starve away they had to be given some small sums by the University for delivering lectures in their specialties. Such a "Lehrauftrag" for algebra was now awarded to Emmy Noether, the first and only salary she was ever to be paid in Göttingen.

She and her work were not on the whole much admired in her native land. She was never even elected to the Göttingen Scientific Society. "It is time that we begin to elect some people of real stature to this society," Hilbert once remarked at a meeting. "Ja, now, how many people of stature have we indeed elected in the past few years?" He looked thoughtfully around at the members. "Only — zero," he said at last. "Only zero!"

A Dutchman, attending one of Emmy Noether's lectures for the first time, remembers that she greeted him, "Ah, another foreigner! I get only foreigners!" But among the foreigners who came to her were van der Waerden from Holland, Artin from Austria, Alexandroff from Russia.

It was Alexandroff who christened her "der Noether", *der* being the definite article which precedes all masculine nouns in German. But he later said: "Her femininity appeared in that gentle and subtle lyricism which lay at the heart of the far-flung but never superficial concerns which she maintained for people, for her profession, and for the interests of all mankind."

She was not a good lecturer and her classes usually numbered no more than five or ten. Once though, she arrived at the appointed hour to find more than a hundred students waiting for her. "You must have the wrong class," she told them. But they began the traditional noisy shuffling of the feet which, in lieu of clapping, preceded and ended each university class. So she went ahead and delivered her lecture to this unusually large number of students. When she finished, a note was passed up to her by one of her regular students who was in the group. "The visitors," it read, "have understood the lecture just as well as any of the regular students."

It was true, she had no pedagogical talents. Her mind was open only to those who were in sympathy with it. Her teaching approach, like her thinking, was wholly conceptual. The German letters which she chalked up on the blackboard were representatives of concepts. It seemed to van der Waerden that "her touching efforts to clarify these, even before she had quite verbalized them . . . had the opposite effect." But of all the new generation in Göttingen, Emmy Noether was to have the greatest influence on the course of mathematics.

While these widening circles of varying mathematical activity were forming themselves around Courant, Landau and Emmy Noether, a group of exceptionally gifted young physicists were gathering around Max Born, who (like Courant, still in his thirties) had become professor of theoretical physics after the war. From the beginning, it was Born's goal to establish at Göttingen a physics institute comparable to Sommerfeld's institute at Munich. When the opportunity arose, he arranged that his best friend, James Franck, join him in Göttingen as professor of experimental physics. It was a stroke reminiscent of Hilbert's bringing Minkowski to the University in 1902. But even before Franck's arrival in 1922, the spectacular series of students who would make their way to Göttingen during the 1920's had begun; and Born's first assistants were Wolfgang Pauli and Werner Heisenberg.

Since the war, the Germans had been barred from most international scientific gatherings; but now it seemed once again that in Göttingen an international congress was permanently in session.

XX

The Infinite!

The highpoint of the mathematical week at Göttingen during the 1920's was the regular session of the Mathematics Club.

The club was a very informal kind of organization without officers, members or dues. Anyone with a doctor's degree could come to meetings, and because of the quality of mathematics at Göttingen it was always "a very high class affair." Sometimes the speaker was a distinguished visitor who reported on his own recent work or that of his students. More often he was a member of the Göttingen circle — professor, docent or student.

The bright young newcomers who saw the famous Hilbert in action for the first time at these events were struck by his slowness in comprehending ideas which they themselves "got" immediately. Often he did not understand the speaker's meaning. The speaker would try to explain. Others would join in. Finally it would seem that everyone present was involved in trying to help Hilbert to understand.

"That I have been able to accomplish anything in mathematics," Hilbert once said to Harald Bohr, "is really due to the fact that I have always found it so difficult. When I read, or when I am told about something, it nearly always seems so difficult, and practically impossible to understand, and then I cannot help wondering if it might not be simpler. And," he added, with his still childlike smile, "on several occasions it has turned out that it really was more simple!"

Some of the young people were irritated by the precious time consumed by Hilbert's questions; others found it fascinating to watch Hilbert's mind in action.

"Scientifically, he did not grasp complicated things at a flash and absorb them. This kind of talent he did not have," Courant explains. "He had to go to the bottom of things."

Hilbert still set high standards of simplicity and clarity for the talks to the Mathematics Club. His guiding rule for the speaker was "only the raisins out of the cake." If computations were complicated, he would interrupt with, "We are not here to check that the sign is right." If an explanation seemed too obvious to him, he would reprove the speaker, "We are not in *tertia*" — *tertia* being the level of the gymnasium in which the student is 12 to 14 years old.

The brutality with which he could dispose of someone who did not meet his standards was well known. There were important mathematicians in Europe and America who dreaded a speech before the Mathematics Club in Göttingen. It seemed now sometimes to Ostrowski that Hilbert was unnecessarily rough on speakers—as if he no longer attended so carefully as in the past to the problem of the superior individual living among lesser individuals.

One young Scandinavian, today highly esteemed, came to Göttingen and spoke about his work — "really important and beautiful and very difficult" in Ostrowski's opinion. Hilbert listened and, when the visitor was through, demanded only, "What is it good for?"

On another occasion he interrupted the speaker with, "My dear colleague, I am very much afraid that you do not know what a differential equation is." Stunned and humiliated, the man turned instantly and left the meeting, going into the next room, which was the Lesezimmer. "You really shouldn't have done that," everyone scolded Hilbert. "But he *doesn't* know what a differential equation is," Hilbert insisted. "Now, you see, he has gone to the Lesezimmer to look it up!"

Still another time the speaker was the young Norbert Wiener. The importance of his talk in Göttingen can be gauged by the fact that many years later in his autobiography he devoted more than a dozen pages to it. After Wiener's talk to the Mathematics Club, everyone hiked up to Der Rohns, as was the custom, and had supper together. Hilbert began in a rambling way during supper to talk about speeches which he had heard during the years he had been at Göttingen.

"The speeches that are given nowadays are so much worse than they used to be. In my time there was an art to giving speeches. People thought a lot about what they wanted to say and their talks were good. But now the young people cannot give good talks any more. It is indeed exceptionally bad here in Göttingen. I guess the worst talks in the whole world are given in Göttingen. This year especially they have been very bad. There have been — no, I have heard no good talks at all. Recently it has been especially bad. But now, this afternoon, there was an exception —"

The young "ex-prodigy" from America prepared himself to accept the compliment.

"This afternoon's talk," Hilbert concluded, "was the worst there ever has been!"

In spite of this remark (which was not reported in the autobiography), Wiener continued to see Hilbert as "the sort of mathematician I [would like] to become, combining tremendous abstract power with a down-to-earth sense of physical reality."

The presence of Klein was still felt in Göttingen during the early twenties, but now like the sunset rather than the high noon-day sun. The editing of the collected works was completed, each paper accompanied by detailed notes on the historical context in which it had originated — a history of the mathematics of his time as well as of his own career. It often seemed to Courant that Klein felt his own life was also completed. He continued to take on projects, such as the editing of his war-time lectures on the history of nineteenth century mathematics, "but with the knowledge that these would have to be finished by others."

When a young mathematician did not immediately follow up a suggestion, Klein dismissed him with, "I am an old man, I can't wait."

Young Norbert Wiener went to pay a call on Klein in the spring of 1925.

"The great man sat in an armchair behind a table, with a rug about his knees. He . . . carried about him an aura of the wisdom of the ages . . . and as he spoke the great names of the past ceased to be the mere shadowy authors of papers and became real human beings. There was a timelessness about him which became a man to whom time no longer had a meaning."

The 1920's were "the beautiful years" when modern physics was developing at an almost magical rate within a triangle which had as its vertices Cambridge, Copenhagen and Göttingen. The 20-year-old Werner Heisenberg, still wearing the khaki shorts of the Youth Movement, came from Munich to Göttingen in 1921. He recalls himself as "much impressed" by the number of young physicists who were interested in the particular problem that was currently interesting Hilbert — "a problem which at that time exceeded by far my own mathematical and physical knowledge." Hilbert had recently returned to his war-time ideas on relativity; and for a while, according to Weyl, hopes ran high in the Hilbert circle of a unified field theory. But, on the whole, it was Hilbert's spirit rather than his person which was felt in physics at this time.

From 1922 on, Hilbert was no longer a physicist. The seminar on the Structure of Matter, which he had instituted with Debye during the war, was now carried on by Born and Franck. The members included at various times during the twenties Heisenberg, Wolfgang Pauli, Robert Oppenheimer, K. T. Compton, Pascual Jordan, Paul Dirac, Linus Pauling, Fritz Houtermans, P. M. S. Blackett among others. Hilbert rarely attended.

His own personal achievement in physics had been a disappointment, "in no way comparable," Weyl later said in summary, "to the mathematical achievement of any single period of his career." The axiomatization of physics, which had been his goal when he first began the joint study with Minkowski, always eluded him.

To Weyl, who himself made important contributions to mathematical physics, it seemed that "the maze of experimental facts which the physicist has to take into account is too manifold, their expansion too fast, and their aspect and relative weight too changeable for the axiomatic method to find a firm enough foothold, except in the thoroughly consolidated parts of our physical knowledge. Men like Einstein or Niels Bohr grope their way in the dark toward their conceptions of general relativity or atomic structure by another type of experience and imagination than those of the mathematician, altough no doubt mathematics is an essential ingredient."

Hilbert's real contribution to physics was to lie in the mathematical methods which he had created in his work on integral equations and in the unification which this work had brought about. When, at the end of 1924, Courant published the first volume of his *Methods of Mathematical Physics*, he placed the name of Hilbert on the title page with his own. This act seemed justified, Courant wrote in the preface, by the fact that much material from Hilbert's papers and lectures had been used as well as by the hope that the book expressed some of Hilbert's spirit, "which had such decisive influence on mathematical research and education."

"Actually it is more than a mere act of dedication that Hilbert's name stands next to that of Courant on the title page," Ewald pointed out in a review of the book for *Naturwissenschaften*. "Hilbert's spirit radiates from the entire book – that elemental spirit, passionately seeking to grasp completely the clear and simple truths, pushing trivialities aside and with masterful clarity establishing connections between the high points of recognition – a spirit that filled generations of searchers with enthusiasm for science."

"Courant-Hilbert," as the book immediately became known, represented a tremendous advance over previous classics of applied mathematics. There had, in fact, really been nothing like it. In the past theoretical physicists had for the most part had to obtain their mathematics from the work of Rayleigh and other physicists. Now they welcomed "Courant-Hilbert."

Hilbert continued to have an assistant to keep him informed on the latest developments in physics. Beginning in 1922, this position was held by Lothar Nordheim, who like all of the other assistants was chosen for Hilbert by Sommerfeld.

Hilbert, in Nordheim's opinion, still had hopes at this time for the achievement of his goal of the axiomatization of physics. To his assistant, however, he was no longer the legendary "great thinker." He was not well. He seemed to live much in the past, had difficulty accepting changes, was prejudiced in many things, his egoism having become more marked. "He could not imagine any greater privilege for a young man than to be his assistant." Nordheim would have preferred a position in Born's institute. Working with Hilbert now in his home, he felt very much out of the mainstream of physics.

But in spite of these signs of apparently early aging, Hilbert continued to maintain his close contacts with youth.

At the same time that Nordheim was coming regularly to Hilbert's house, another young man was also a frequent visitor. John von Neumann had studied in Berlin with Erhard Schmidt, Hilbert's former student who, at the beginning of the century, had so significantly forwarded the Hilbert work on integral equations. He was a young man who was in one respect at least the exact opposite of Hilbert. Whereas Hilbert was "slow to understand," von Neumann was equipped with "the fastest mind I ever met," according to Nordheim. He frequently expressed his opinion that the mathematical powers decline after the age of 26, but that a certain prosaic shrewdness developing from experience manages to compensate for this gradual loss. (During his own life, he slowly raised the limiting age.)

Von Neumann was 21 in 1924, deeply interested in Hilbert's approach to physics and also in his ideas on proof theory. The two mathematicians, more than forty years apart in age, spent long hours together in Hilbert's garden or in his study.

Hilbert's real collaborator during these days, however, was Bernays. To some people it seemed that he was even exploiting his logic assistant. Bernays was no young student but a man in his middle thirties, a mature mathematician. As Hilbert's assistant, he received a salary and, having habilitated shortly after his arrival in Göttingen, also received fees from the students who attended his lectures. He could live on what he received, but certainly he could not marry.

Hilbert was very opposed to marriage for young scientists anyway. He felt that it kept them from fulfilling their obligations to science. Later, when Wilhelm Ackermann, with whom he had worked and collaborated on a book, married, Hilbert was very angry. He refused to do anything more to further Ackermann's career; and as a result, not obtaining a university position, the gifted young logician had to take a job teaching in a high school. When, sometime later, Hilbert heard that the Ackermanns were expecting a child, he was delighted.

"Oh, that is wonderful!" he said. "That is wonderful news for me. Because if this man is so crazy that he gets married and then even has a child, it completely relieves me from having to do anything for such a crazy man!"

In addition to preparing his own lectures, Bernays helped Hilbert prepare his lectures, accompanied him to class and often took over the teaching for part of the hour, supervised Hilbert's students who were working for the doctoral degree, studied and digested the literature necessary for their work, and did a great deal of writing on their joint book, which was to be entitled *Grundlagen der Mathematik*. In Bernays, Hilbert had found someone as interested in the foundations of mathematics as he was. He had no compunction about working his assistant as hard as he worked himself. "Genius is industry," he liked to tell his students and his assistants, quoting Lichtenberg. He himself was, as Weyl later recalled, "enormously industrious."

The two men sometimes, however, got into rather violent arguments over the subject of foundations. Bernays attributes the emotional quality of these arguments to a fundamental "opposition" in Hilbert's feelings about mathematics.

"For Hilbert's program," he explains, "experiences out of the early part of his scientific career (in fact, even out of his student days) had considerable significance; namely, his resistance to Kronecker's tendency to restrict mathematical methods and, particularly, set theory. Under the influence of the discovery of the antinomies in set theory, Hilbert temporarily thought that Kronecker had probably been right there. But soon he changed his mind. Now it became his goal, one might say, to do battle with Kronecker with his own weapons of finiteness by means of a modified conception of mathematics . . . .

"In addition, two other motives were in opposition to each other — both strong tendencies in Hilbert's way of thinking. On one side, he was convinced of the soundness of existing mathematics; on the other side, he had — philosophically — a strong scepticism."

An example was Hilbert's attitude toward the question of the solvability of every definite mathematical problem. At Paris he had spoken in ringing tones of the axiom of the solvability of every problem, "the conviction which every mathematician shares, although it has not yet been supported by proof." He was convinced that in mathematics at least "there is no *ignorabimus*." Yet, at Zürich, he listed among the epistemological questions which he felt should be investigated the question of the solvability in principle of every mathematical question.

"The problem for Hilbert," Bernays explains, "was to bring together these opposing tendencies, and he thought that he could do this through the method of formalizing mathematics."

Bernays did not always agree with Hilbert about their program, but he appreciated the fact that, passionate though Hilbert was in his disputation, he never held it against his assistant personally when he took the opposite side.

After their work was finished, Hilbert and Bernays often argued about politics. Hilbert enjoyed expressing his views on the subject in extreme and paradoxical ways.

Although he was generally considered conservative, he surprised everybody by proposing Käthe Kollwitz, who was known to be very strongly oriented to the left, for the Star of the Order of Merit, Peace Class. She had become one of the great women artists of all time. ("I have never seen such a drawing by the hand of a woman," the sculptor Constantin Meunier said.) Her subject matter reflected her feeling for the sufferings of humanity.

"Of course what she draws is horrible to look at," Hilbert told his fellow wearers of the Star. "But when we were young in Königsberg and used to dance, she was one of the first girls to dance without her corset!"

In spite of his conservative background, Hilbert was always a liberal in the sense that he never considered himself bound to any certain political view. In his arguments with his assistant, he often criticized "liberals" for seeing things as they wished them to be and not as they were.

"Sometimes," he said, "it happens that a man's circle of horizon becomes smaller and smaller, and as the radius approaches zero it concentrates on one point. And then that becomes his point of view."

He liked to remind his younger assistant: "Mankind is always the constant."

Music often brought peace after the arguments, logical or political. Bernays loved music and had played "four hands" with Hurwitz when he was in Zürich. He was impressed by how much Hilbert's musical knowledge and appreciation had developed over the years as a result of his love for his phonograph, a new model of which was still being supplied regularly to him by the manufacturer. Now he was a member of a group of professors and their wives who attended concerts together in Göttingen and traveled to Leipzig or Hannover for special musical events.

Hilbert sometimes seemed to have very little appreciation of the arts other than music. Yet he was drawn to literature and, as Courant says, "wanted to be aware." He appreciated Goethe and Homer, but he insisted on action in novels. One of his "flames" once set out to educate him in literature. She began by giving him a historical novel about the Swiss civil wars, quite a bloody tale. Hilbert shortly returned it. "If I am given a book to read," he said, "it should be one in which something really happens. Describing the soul and the variations of mood – that I can do for myself!"

There is one story about his attitude toward literature which also reveals a great deal about his feeling for mathematics. It seems that there was a mathematician who had become a novelist. "Why did he do that?" people in Göttingen marvelled. "How can a man who was a mathematician write novels?" "But that is completely simple," Hilbert said. "He did not have enough imagination for mathematics, but he had enough for novels."

The current creation of Hilbert's own mathematical imagination was his proof theory. At Zürich, in 1917, he had announced the general idea and aims of the theory – but not the means of investigation. "For indeed," Bernays was later to comment, "the theory was not to rely on the current mathematical methods." In the first communication on the new theory (the attack on Brouwer and Weyl in Hamburg in 1922), Hilbert had argued that mathematicians could regain elementary objectivity by formalizing the statements and proofs of mathematics in the language of symbolic logic and then taking the represented formulas and proofs directly as objects for study. That same year, at Leipzig, he had added further refinements which reduced the problem of proving the consistency of a formalized domain of arithmetic – the task which he had set for the new century in Paris in 1900.

"Thus it seemed," Bernays later wrote, "that carrying out proof theory was only a matter of mathematical technique."

On the occasion of a celebration at Münster in honor of Weierstrass, Hilbert chose to talk "On the Infinite." He felt that the occasion was appropriate for the fullest exposition of his program of Formalism up to that time. The analysis of Weierstrass and the concept of the infinite as it appeared in the work of Cantor had been prime targets for Kronecker. In the current program of Brouwer, many of the achievements of Weierstrass and Cantor would be among the sacrifices required.

Hilbert was not at all well at the time of his talk in Münster. Recently it had become clear that the deterioration noted by Nordheim was not that of age alone, but the nature of the illness was still undetermined. In spite of his poor health, however, Hilbert spoke as enthusiastically and optimistically as ever.

He began his talk by pointing out that the present "happy state of affairs" in analysis was entirely due to Weierstrass and his penetrating critique of its methods. And yet — disputes about the foundations of analysis did continue up to the present day. This was, in his opinion, because the meaning of *the infinite*, as that concept was used in mathematics, had not yet been completely clarified.

The infinite was nowhere to be found in reality; yet it existed in a very real sense, in his opinion, as an "over-all negation." From time immemorial the idea of the infinite had stirred men's emotions as no other subject. Therefore, he felt, the definitive clarification of its nature went far beyond the sphere of specialized scientific interest: it was needed for the dignity of the human intellect itself!

The deepest insight into the nature of the infinite to date had been obtained by a theory which came closer to a general philosophical way of thinking than to mathematics. This theory, created by Georg Cantor, was set theory.

"It is, I think, the finest product of mathematical genius," Hilbert said, "and one of the supreme achievements of purely intellectual human activity."

But it was in Cantor's set theory, simply as a result of employing definitions and deductive methods which had become customary in mathematics, that the catastrophic antinomies had begun to appear.

". . . the present state of affairs . . . is intolerable. Just think, the definitions and deductive methods which everyone learns, teaches, and uses in mathematics, the paragon of truth and certitude, lead to absurdities! If mathematical thinking is defective, where are we to find truth and certitude?"

There was, however, "a completely satisfactory way of avoiding the paradoxes of set theory without betraying our science." Mathematicians must establish throughout mathematics the same certitude for their deductions as exists in the ordinary arithmetic of whole numbers, "which no one doubts and where contradictions and paradoxes arise only through our own carelessness."

But if men were to remain within the domain of such purely intuitive and finitary statements — as they must — they would have to have, as a rule, more complicated logical laws. The logical laws which Aristotle had taught and which men had used since they began to think, would not hold.

"We could, of course, develop logical laws which do hold for the domain of finitary statements. But . . . we do not want to give up the use of the simple laws of aristotelian logic . . . . What then are we to do?

"Let us remember that *we are mathematicians* and that as mathematicians we have often been in precarious situations from which we have been rescued by the ingenious method of ideal elements . . . . Similarly, to preserve the simple formal rules of aristotelian logic, we must *supplement the finitary statements with ideal statements.*"

Mathematics, under this view, would become a stock of two kinds of formulas: first, those to which the meaningful communications correspond and, secondly, other formulas which signify nothing but which are the ideal structures of the theory.

"But in our general joy over this achievement, and in our particular joy over finding that indispensable tool, the logical calculus, already developed without any effort on our part, we must not forget the essential condition of the method of ideal elements — *a proof of consistency.*"

For the extension of a domain by the addition of ideal elements is legitimate only if the extension does not cause contradictions to appear.

This problem of consistency could be "easily handled." It was possible, in his opinion, to obtain in a purely intuitive and finitary way — the way in which truths are obtained in elementary number theory — the insights which would guarantee the validity of the mathematical apparatus. Then the test of the theory would be its ability to solve old problems, for the solution of which it had not been expressly designed. He cited as an example of such a problem the continuum hypothesis of Cantor, which he had listed as the first of the Paris Problems. He now devoted the last part of his talk to sketching an attack on this famous problem.

"No one," he promised his fellow mathematicians, "will drive us out of this paradise that Cantor has created for us!"

XXI

Borrowed Time

On a soft warm evening in June 1925, Felix Klein died.

Everyone at Göttingen had long been prepared for Klein's death.

"But the event after it happened touched us all deeply and affected us painfully," Hilbert said in a little speech to his colleagues the next morning. "Up until yesterday Felix Klein was still with us, we could pay him a visit, we could get his advice, we could see how highly interested he was in us. But that is now all over."

Everything they saw around them in Göttingen was the work of Klein, the collection of mathematical models in the adjoining corridor, the Lesezimmer with all the books on open shelves, the numerous technical institutes that had grown up around the University, the easy relation they had with the education ministry, the many important people from business and industry who were interested in them . . . . They had lost "a great spirit, a strong will, and a noble character."

An era had come to an end.

A few months later at a memorial meeting of the Göttingen Scientific Society, Courant recalled the dramatic story of the great Felix: the meager beginnings, the spectacular successes ("If today we are able to build on the work of Riemann, it is thanks to Klein."), the tragic breakdown, and then – "the wonderful turning point" – the seemingly broken man who had lived another 43 years and displayed the most varied sides as researcher, teacher, organizer and administrator.

And yet Klein's life had not been without its inner tragedy. The power of synthesis had been granted to him to an extraordinary degree. The other great mathematical power of analysis had been to a certain extent withheld. His ability to bring together the most distant, abstract parts of mathematics had been remarkable, but the sense for the formulation of an individual problem and the absorption in it had been lacking. "He was like a flier who, soaring high over the world, discovers and looks over new fields . . . but cannot land his plane in order to take actual possession, to plow and to harvest." Perhaps Klein had himself been unaware of this deep schism but, in Courant's opinion, it had been one of the causes of the decisive breakdown during his competition with Poincaré. Certainly he had perceived "that his most splendid scientific creations were fundamentally gigantic sketches, the completion of which he had to leave to other hands."

Sometimes he had failed to preserve purely human relationships. "Many who knew him only as an organizer . . . found him too harsh and violent, so he produced much opposition to his ideas . . . which a gentler hand would easily have overcome." Yet his nearest relatives and colleagues and the great majority of his students had known always that behind the relentlessly naive drive, a good human being stood.

They placed on his grave a simple inscription: "Felix Klein, A Friend, Sincere and Constant."

The same year that Klein died Runge retired; his place was taken by Gustav Herglotz.

The ailing Hilbert's condition steadily worsened. In the fall of 1925 it was at last recognized that he was suffering from pernicious anemia. The disease, which was generally considered fatal, had gone so long undetected because the first symptoms, occurring in someone of his age, seemed merely an early failing of powers. Now the doctors gave him at best a few months, maybe even weeks.

In spite of what the doctors said, Hilbert remained optimistic about his condition. He insisted that he actually did not have pernicious anemia — it was some other, less serious disease which merely had all the same symptoms.

It was Hilbert's sheer luck that earlier in the year 1925 G. H. Whipple and F. S. Robscheit-Robbins had discovered the beneficial effects of raw liver on blood regeneration, and by 1926 their work was being applied to the treatment of pernicious anemia by G. R. Minot in America. A pharmacologist friend in Göttingen now chanced to read about Minot's work in the *Journal of the American Medical Association* and showed the article to Hilbert. In addition to describing the new treatment — still, it stressed, in a highly experimental state — the article also described vividly the mortal seriousness of "P. A." But Hilbert reading it completely ignored all the distressing details. He concentrated only upon the hopes raised by Minot's work.

Mrs. Landau was the daughter of Paul Ehrlich, who had discovered salvarsan, the "magic bullet" treatment for syphilis; and she had many contacts in the medical world. With the help of Courant, she drafted a long telegram to Minot, who was at Harvard. "It was the longest telegram I ever sent," Courant says. At the same time another telegram went to Oliver Kellogg who, in 1902, had been the first Hilbert student to write his doctoral dissertation on integral equations. Now a mathematics professor at Harvard, Kellogg rallied support among the mathematicians for the request from Göttingen.

At first Minot and his associates were not too receptive. They had very little of the treatment substance, which would have to be administered to the patient for the rest of his life. People were dying of pernicious anemia within a few miles of Harvard University . . . .

George Birkhoff, the leading American mathematician and also a Harvard mathematics professor, had recently seen a play in which a doctor was able to save but 10 men. How should he choose the 10? "On the basis of their value to mankind" was the playwright's answer. In conversation with Minot, Birkhoff quoted *The Doctor's Dilemma* of George Bernard Shaw.

Mathematics makes mathematicians persistent men. Minot gave in. Instructions were wired to the pharmacologist in Göttingen for concocting large amounts of raw liver which would serve for treatment until the more concentrated experimental substance arrived from the United States. E. U. Condon, visiting Göttingen in the summer of 1926, heard Hilbert complaining that he would rather die than eat that much raw liver.

Eventually, though, Minot's preparation arrived.

At this late stage it was probably not possible to reverse completely the progress of the disease; however, everyone in Göttingen noticed that Hilbert's condition began to improve almost immediately. All during his illness he had continued to work to the best of his ability – even turning his dining room into a lecture hall when he was not well enough to go to the University. Now when a former student inquired after his health, he replied firmly, "That illness – well, it no longer *exists*."

The two years, beginning in 1925, were the "Wunderjahre" of what was known in Göttingen as "boy physics" because so many of the great discoveries were being made by physicists still in their twenties. Early in 1925 Heisenberg came to Born with the seemingly weird mathematics that had developed in a new theory of quantum mechanics which he had created. Heisenberg thought that this was the one thing that still had to be corrected in his theory. In actuality, it was his great discovery. Born promptly identified the weird mathematics as matrix algebra, the germ of which had existed in the quaternions developed by William Rowan Hamilton more than three-quarters of a century before.

In matrix algebra multiplication is not commutative: $a \times b$ may not equal $b \times a$ but something entirely different. Prior to Heisenberg's work, matrices had rarely been used by physicists, although one exception had been Born's earlier work on the lattice theory of crystals. But now even Born had to consult his old friend Otto Toeplitz about certain properties of matrices and considered himself fortunate to obtain as his assistant Pascual Jordan, whom he just happened to meet when Jordan overheard him talking about matrices to a companion in a train compartment and proceeded to introduce himself. Jordan had been one of Courant's assistants in the preparation of Courant-Hilbert and was therefore very familiar with matrix algebra.

The Heisenberg paper was followed just 60 days later by the great Born-Jordan paper, which provided the necessarily rigorous mathematical foundation for the new matrix mechanics. The next year saw the publication of Born's famous statistical interpretation, for which he later received the Nobel Prize.

Hilbert never went as deeply into quantum mechanics as he had gone into relativity, but he still demanded that his physics assistant teach him the new theory.

"Generally he tried to give a course on what he was learning," says Nordheim. "He was a person for whom it was difficult to understand others. He always had to work things through for himself. That seemed to be his only way of really understanding. So when there was a new development, he tried to give a course on it. Usually this also contained some old material, for nothing grows up entirely of itself. For the new parts we had to make drafts. After that he would try to put the ideas into his own words."

In the spring of 1926 Hilbert announced his first lectures on quantum mechanics. Nordheim recalls how he had to extract "rather laboriously" the essence from the papers of Born and his collaborators for Hilbert, who was still at this time not well.

"Of course, he knew a lot about matrix algebra and differential equations and so on, and all of these things are the mathematical tools of quantum mechanics. In this respect my job was made easier. I went to his home two or three times a week, as required, and we discussed the general situation. Then he would ask for writeups on specific points or the development of formulas on particular applications. The next time we would talk these over — whether everything was correct and understood."

The matrix mechanics of Heisenberg was followed in short order by the wave mechanics of Erwin Schrödinger. The two papers, although they were on the same subject and led to the same results, astonished physicists; for, as one of them marvelled, "they started from entirely different physical assumptions, used entirely different mathematical methods, and seemed to have nothing to do with each other."

The equivalence of Heisenberg's and Schrödinger's theories, however, was soon established.

The whole development gave Hilbert "a great laugh," according to Condon:

". . . when [Born and Heisenberg and the Göttingen theoretical physicists] first discovered matrix mechanics they were having, of course, the same kind of trouble that everybody else had in trying to solve problems and to manipulate and to really do things with matrices. So they had gone to Hilbert for help and Hilbert said the only times he had ever had anything to do with matrices was when they came up as a sort of by-product of the eigenvalues of the boundary-value problem of a differential equation. So if you look for the differential equation which has these matrices you can probably do more with that. They had thought it was a goofy idea and that Hilbert didn't know what he was talking about. So he was having a lot of fun pointing out to them that they could have discovered Schrödinger's wave mechanics six months earlier if they had paid a little more attention to him."

As a result of his almost miraculous recovery Hilbert lived to see what has been called "one of the most dramatic anticipations in the history of mathematical physics."

The Courant-Hilbert book on mathematical methods of physics, which had appeared at the end of 1924, before both Heisenberg's and Schrödinger's work, instead of being outdated by the new discoveries, seemed to have been written expressly for the physicists who now had to deal with them. Hilbert's own work at the beginning of the century on integral equations, the theory of eigenfunctions and eigenvalues of 1903–04 and the theory of infinitely many variables of 1905–06, turned out to be the appropriate mathematics for quantum mechanics (as was first established by Born in a joint paper with Heisenberg and Jordan).

"Indirectly Hilbert exerted the strongest influence on the development of quantum mechanics in Göttingen," Heisenberg was later to write. "This influence can be fully recognized only by one who studied in Göttingen during the twenties. Hilbert and his colleagues had created there an atmosphere of mathematics, and all the younger mathematicians were so trained in the thought processes of the Hilbert theory of integral equations and linear algebra that each project which belonged in this field could develop better in Göttingen than in any other place. It was an especially fortunate coincidence that the mathematical methods of quantum mechanics turned out to be a direct application of Hilbert's theory of integral equations . . . ."

To Hilbert himself this was yet another example of that pre-established harmony which seemed to him almost the embodiment and realization of mathematical thought.

"I developed my theory of infinitely many variables from purely mathematical interests," he marvelled, "and even called it 'spectral analysis' without any presentiment that it would later find an application to the actual spectrum of physics!"

What happened next was also impressive to Hilbert, for it underlined the continuity of mathematical effort. Hilbert's theory of infinitely many variables – which had become known as "Hilbert Space" theory – now turned out to be in several respects not quite equal to the task of handling quantum mechanics. At this point young John von Neumann, inspired by Erhard Schmidt, formulated Hilbert's concept of a quadratic form more abstractly so that the extended Hilbert theory was able to meet completely the needs of the physicists.

Hilbert's last publication in physics was a collaboration with Nordheim and von Neumann on the axiomatic foundations of quantum mechanics, in which, although he did almost none of the work, the spirit was quite definitely that of Hilbert. While the effort turned out to be not mathematically rigorous, it served to introduce von Neumann to quantum mechanics and inspired him to create his later famous analysis of the foundations of the subject.

In 1927 Nordheim left Göttingen and Eugene Wigner became Hilbert's special assistant for physics. He recalls that he saw Hilbert "only about five times." When Wigner left in 1928, his place was taken by a mathematics student named Arnold Schmidt and the position became a second assistantship in logic. Hearing a lecture by Schrödinger on the new physics in 1928 or 1929, Hilbert grumbled to his former student Paul Funk: "I don't see how anybody understands what is happening in physics today. Even I don't understand much which I would like to learn from physics books. But with me, if I don't understand something, then I go to the telephone and call up Debye or Born, and they come and explain it to me. And then I understand it — but what do other people do?"

He himself was still, after his illness, deep in the work on the foundations of mathematics.

The enthusiasm for Brouwer's Intuitionism had definitely begun to wane. Brouwer came to Göttingen to deliver a talk on his ideas to the Mathematics Club.

"You say that we can't *know* whether in the decimal representation of π ten 9's occur in succession," someone objected after Brouwer finished. "Maybe we can't know — but God knows!"

To this Brouwer replied dryly, "I do not have a pipeline to God."

After a lively discussion Hilbert finally stood up.

"With your methods," he said to Brouwer, "most of the results of modern mathematics would have to be abandoned, and to me the important thing is not to get fewer results but to get more results."

He sat down to enthusiastic applause.

The feeling of most mathematicians has been informally expressed by Hans Lewy, who as a Privatdozent was present at Brouwer's talk in Göttingen:

"It seems that there are some mathematicians who lack a sense of humor or have an over-swollen conscience. What Hilbert expressed there seems exactly right to me. If we have to go through so much trouble as Brouwer says, then nobody will want to be a mathematician any more. After all, it is a human activity. Until Brouwer can produce a contradiction in classical mathematics, nobody is going to listen to him.

"That is the way, in my opinion, that logic has developed. One has accepted principles until such time as one notices that they may lead to contradiction and then he has modified them. I think this is the way it will always be. There may be lots of contradictions hidden somewhere; and as soon as they appear, all mathematicians will wish to have them eliminated. But until then we will continue to accept those principles that advance us most speedily."

Hilbert's program, however, also received its share of criticism. Some mathematicians objected that in his formalism he had reduced their science to "a meaningless game played with meaningless marks on paper." But to those familiar with Hilbert's work this criticism did not seem valid.

". . . is it really credible that this is a fair account of Hilbert's view," Hardy demanded, "the view of the man who has probably added to the structure of significant mathematics a richer and more beautiful aggregate of theorems than any other mathematician of his time? I can believe that Hilbert's philosophy is as inadequate as you please, but not that an ambitious mathematical theory which he has elaborated is trivial or ridiculous. It is impossible to suppose that Hilbert denies the significance and reality of mathematical concepts, and we have the best of reasons for refusing to believe it: 'The axioms and demonstrable theorems,' he says himself, 'which arise in our formalistic game, are the images of the ideas which form the subject-matter of ordinary mathematics.'"

By 1927 Hilbert was well enough to go again to Hamburg "to round out and develop my thoughts on the foundations of mathematics, which I expounded here one day five years ago and which since then have kept me most actively occupied." His goal was still to remove "once and for all" any question as to the soundness of the foundations of mathematics. "I believe," he said, "I can attain this goal completely with my proof theory, even though a great deal of work must still be done before it is fully developed."

In the course of his talk, he took up various criticisms of his program, "all of which I consider just as unfair as can be." He went back even as far as Poincaré's remarks on the Heidelberg talk. "Regrettably, Poincaré, the mathematician who in his generation was the richest in ideas and the most fertile, had a decided prejudice against Cantor's theory that kept him from forming a just opinion of Cantor's magnificent conceptions." As for the most recent investigations, of which the program advanced by Brouwer formed the greater part, "the fact that research on foundations has again come to attract such lively appreciation and interest certainly gives me the greatest pleasure. When I reflect on the content and the results of these investigations, however, I cannot for the most part agree with their tendency; I feel, rather, that they are to a large extent behind the times, as if they came from a period when Cantor's majestic world of ideas had not yet been discovered."

The whole talk had a strongly polemical quality: "Not even the sketch of my proof of Cantor's continuum hypothesis has remained uncriticized!" Hilbert complained, and took up this proof again at length.

The formula game "which Brouwer so deprecates," he pointed out, enabled mathematicians to express the entire thought-content of the science of mathematics in a uniform manner and develop it in such a way that, at the same time, the interconnections between the individual propositions and the facts become clear. It had, besides its mathematical value, an important general philosophical significance.

"For this formula game is carried out according to certain definite rules, in which the *technique of our thinking* is expressed. These rules form a closed system that can be discovered and definitively stated. The fundamental idea of my proof theory is none other than to describe the activity of our understanding, to make a protocol of the rules according to which our thinking actually proceeds .... If any totality of observations and phenomena deserves to be made the object of a serious and thorough investigation, it is this one – since, after all, it is a part of the task of science to liberate us from arbitrariness, sentiment and habit and to protect us from the subjectivism that already made itself felt in Kronecker's views and, it seems to me, finds its culmination in Intuitionism ...."

It was true, Hilbert conceded, that the consistency proof of the formalized arithmetic which would "determine the effective scope of proof theory and in general constitute its core" was not yet at hand. But, as he concluded his address, he was thoroughly optimistic: such a proof would soon be produced.

"Already at this time I should like to assert what the final outcome will be: mathematics is a presuppositionless science. To found it I do not need God, as does Kronecker, or the assumption of a special faculty of our understanding attuned to the principle of mathematical induction, as does Poincaré, or the primal intuition of Brouwer, or, finally, as do Russell and Whitehead, axioms of infinity, reducibility, or completeness ...."

When Hilbert finished, Hermann Weyl rose to make a few remarks. Weyl's love for his old teacher had not been affected by the five years of controversy. Also his enthusiasm for Brouwer's ideas had abated. He nevertheless felt that at this point he should defend Brouwer:

"Brouwer was first to see exactly and in full measure how [mathematics] had in fact everywhere far exceeded the limits of contentual thought. I believe that we are all indebted to him for this recognition of the limits of contentual thought. In the contentual considerations that are intended to establish the consistency of formalized mathematics, Hilbert fully respects these limits, and he does so as a matter of course; we are really not dealing with artificial prohibitions by any means. Accordingly, it does not seem strange to me that Brouwer's ideas have found a following; his position resulted of necessity from a thesis shared by all mathematicians before Hilbert proposed his formal approach and forms a new, indubitable fundamental logical insight that even Hilbert acknowledges.

"That from this point of view only a part, perhaps only a wretched part of classical mathematics is tenable is a bitter and inevitable fact. Hilbert could not bear this mutilation. And it is again a different matter that he succeeded in saving classical mathematics by a radical reinterpretation of its meaning without reducing its inventory, namely by formalizing it, thus transforming it in principle from a system of intuitive results into a game with formulas that proceeds according to fixed rules.

"Let me now by all means acknowledge the immense significance and scope of this step of Hilbert's, which evidently was made necessary by the pressure of circumstances. All of us who witnessed this development are full of admiration for the genius and steadfastness with which Hilbert, through his proof theory of formalized mathematics, crowned his axiomatic life work. And, I am very glad to confirm, that there is nothing that separates me from Hilbert in the epistemological appraisal of the new situation thus created."

In contrast to Weyl, Brouwer had become, like Kronecker, a fanatic in the service of his cause. He looked upon Hilbert as "my enemy," and once left a house in Amsterdam where he was a guest when van der Waerden, who was also a guest, referred to Hilbert and Courant as his friends.

The ill-feeling was undoubtedly intensified by the fact that circumstances placed Hilbert constantly in opposition to Brouwer.

Both men were on the editorial staff of the *Annalen*. Hilbert was one of the three principal editors, a position which he had held since 1902; Brouwer, a member of the seven-man editorial board. At about this time Brouwer began to insist that all papers by Dutch mathematicians and all papers on topology be submitted directly to him. Everyone objected, especially Dutch topologists, since it was well known that when a paper got into Brouwer's hands, it did not get out for several years. Although personally unaffected, Hilbert was repelled by Brouwer's dictatorial demands. When he had been in good health, he had been confident of his ability to protect the integrity of the *Annalen*. Since his illness, however, he had begun to fear that if anything happened to him, Brouwer would take over the journal to the detriment of mathematics. So he now called together his friends to devise a way of removing Brouwer from the editorial board.

Carathéodory, who was himself a member of the board, came up with a solution. Since Brouwer alone could not be asked to resign, the entire seven-man board should be dismissed. Hilbert promptly acted. The change is reflected in the covers of Vol. 100 of the *Annalen* and Vol. 101, on which only the names of Hilbert, Hecke and Blumenthal remain.

(It should be mentioned that Einstein, disturbed by the controversy, resigned from his position as one of the three principal editors. "What is this frog and mouse battle among the mathematicians?" he asked a friend.)

Hilbert and Brouwer were then placed in opposition by another circumstance.

Since the war, the German mathematicians had not been invited to any international meetings. In 1928, however, the Italians, planning the first official International Congress since 1912, determined to make it truly international. Once again, invitations were sent to all German schools and mathematical organizations. Many Germans did not want to accept. The leader of this group was a professor at the University of Berlin named Ludwig Bieberbach. In his opposition to accepting the invitation of the Italians, he was seconded by Brouwer, who although Dutch was an ardent German nationalist. In the spring of 1928 Bieberbach sent a letter to all German secondary schools and universities urging them to boycott the congress at Bologna. Hilbert responded by sending out a letter of his own:

"We are convinced that pursuing Herr Bieberbach's way will bring misfortune to German science and will expose us all to justifiable criticism from well disposed sides . . . . The Italian colleagues have troubled themselves with the greatest idealism and expense in time and effort . . . . It appears under the present circumstances a command of rectitude and the most elementary courtesy to take a friendly attitude toward the Congress."

In August, although suffering from a recurrence of his illness, Hilbert personally led a delegation of 67 mathematicians to the Congress. At the opening session, as the Germans came into an international meeting for the first time since the war, the delegates saw a familiar figure, more frail than they remembered, marching at their head. For a few minutes there was not a sound in the hall. Then, spontaneously, every person present rose and applauded.

"It makes me very happy," Hilbert told them in the familiar accents, "that after a long, hard time all the mathematicians of the world are represented here. This is as it should be and as it must be for the prosperity of our beloved science.

"Let us consider that we as mathematicians stand on the highest pinnacle of the cultivation of the exact sciences. We have no other choice than to assume this highest place, because all limits, especially national ones, are contrary to the nature of mathematics. It is a complete misunderstanding of our science to construct differences according to peoples and races, and the reasons for which this has been done are very shabby ones.

"Mathematics knows no races . . . . For mathematics, the whole cultural world is a single country."

Hilbert's scientific paper for presentation at the Congress concerned again the fundamental problems of mathematics. Recently there had been signs that his expectation that carrying out proof theory was only a matter of mathematical technique was perhaps overly optimistic. The first attempt at a substantial consistency proof (in Ackermann's dissertation) had required an essential restriction of the formal system not in the original plan. Similarly, in a paper of von Neumann's, the consistency proof, following Hilbert's line of reasoning, had not applied to the full system. Now, though, Ackermann's proof had been revised and simplified; and it seemed at the moment that the consistency of formalized number theory, at least, had been proved.

Hilbert now added to the problem of consistency another problem, that of the *completeness* of the formal system.

When Hilbert went to pay his hotel bill, he was informed that it had already been paid for him by the committee in charge of planning the Congress.

"Ah, if I had only known that," he said, "I would have eaten a great deal more."

The Hilbert career was almost over.

The year after the Bologna Congress, he was permitted to see what Felix Klein had not lived to see — the dedication of a handsome building to house the Mathematical Institute of Göttingen.

The new institute had been made possible by Courant's friendship with the Bohr brothers and the entrée which they had provided to the Rockefeller Foundation. The funds given by the Foundation were then matched by the German government. Thus the Institute was a joint project of German and American money and effort.

"There will never be another institute like this!" Hilbert exulted. "For to have another such institute, there would have to be another Courant — and there can never be another Courant!"

XXII

Logic and the Understanding of Nature

The mandatory age of retirement for a professor was 68, an age Hilbert would attain on January 23, 1930; in Göttingen a bittersweet feeling of anticipation and regret was in the air.

During the winter semester 1929–30, Hilbert delivered his "Farewell to Teaching." For his subject he went back to the foundations of his fame and lectured for almost the first time in 40 years on the invariants. Professors crowded into the lecture hall with the students. A street was named Hilbert Strasse. "A street named after you!" Mrs. Hilbert exclaimed. "Isn't that a *nice* idea, David?" Hilbert shrugged. "The idea, no, but the execution, ah – that is nice. Klein had to wait until he was dead to have a street named after him!"

He saw another student through the doctoral. Appropriately this was Haskell Curry, an American. Curry had little contact with Hilbert though. He remembers he came to class on a warm spring day wearing a fur-lined coat. He was always accompanied by Bernays, who sometimes had to step in and lecture for a bit. Curry had most of his conferences with Bernays; but since Bernays was not a full professor, he had to take the final examination with Hilbert.

"I rather enjoyed my final examination with him . . . . He did not ask me any questions having to do with logic, but only to do with general mathematics. One question was about the uniformization of algebraic functions. It happened that I had just had a course on that subject with Professor Osgood at Harvard. Although it was a way off from my special field, I gave as precise an answer as anyone would expect in a field so far removed from the candidate's specialty; he was quite impressed with me and turned to me and said, 'Where did you learn that?' Although he seemed rather frail, he was razor-sharp and alert."

As the time of the retirement approached, the choice of a successor was discussed. There was, it was generally agreed, but one possible choice. If Courant had shown himself to be the Klein of the new generation, Weyl was the Hilbert.

A decade before, Weyl had refused an offer from Göttingen because of the uncertainties of life in Germany after the war. Hilbert said, "It is easy to call Weyl but hard to get him." This time Weyl again had difficulty in making up his mind. He had recently returned from England; and the pessimism which he found expressed in the newspapers and letters which had piled up on his desk during his absence filled him with apprehension about returning to Germany. He was also concerned that he might not be the right choice for Göttingen at this time. By now he was 45 years old. He knew he was close to the end of his creative period. Perhaps the Institute should get someone like young Emil Artin, "from whom great results can still be expected." Yet he was tempted. He loved and revered Hilbert – the Pied Piper who had led all the young rats down into the deep river of mathematics. He knew that he was more completely in the mathematical-physical tradition of Göttingen than Artin. He would enjoy the opportunity of working with Courant, Born and Franck. The situation in Germany seemed to be improving. The Dawes Plan had helped to relieve the economic problems. The lunatic fringe which muttered about the "Jewish physics" of his friend Einstein still seemed on the fringe. In the end, this time, Weyl wired his acceptance.

"I don't have to tell you with how much joy and how much pride I was filled to be called as your successor," he wrote to Hilbert. ". . . I am looking forward most eagerly to working with the colleagues you have gathered around yourself, you to whom the mathematical-scientific faculty owes its strength and harmony." The dark clouds that hung over Germany might not disappear quickly. "But I hope it will be granted to me to live many happy years near you . . . . Please do not be angry about my tardiness in accepting."

The Göttingen which greeted Weyl in the spring of 1930 was at the height of its new glory. More than ever before, it could be said that an international congress of mathematicians was perpetually in session in the quiet little town with its linden-lined streets and the solid respectable, now old "Jugendstil" houses. On the outskirts a series of scientific industries and laboratories surrounded the city like another wall. The Mathematical Institute was housed in its handsome new building, the Lesezimmer a long well-lighted library. *Extra Gottingen non est vita.* The Latin motto was still blazoned on the wall of the Ratskeller. In the sunshine outside, students and professors sat at small tables and argued about politics, love and science. The little goose girl gazed tranquilly down into her fountain. Weyl, returning to the beloved town of his college years, must have agreed. Away from Göttingen there was no life.

Of all the honors being showered on Hilbert during the retirement year, the one that seemed to please him most came from his native city. The Königsberg town council voted to present its famous son with "honorary citizenship." The presentation was scheduled to be made in the fall at the meeting of the Society of German Scientists and Physicians, which was being held that year in Königsberg.

Hilbert gave considerable thought to the selection of a topic for his acceptance address. It must be something of wide and general interest. In Königsberg, the birthplace of Kant, it must be philosophical in tone. It must also be a fitting conclusion to the career that had begun long ago at the university in Königsberg. When he thought of the university, he remembered the statue of Kant on the grounds and the laconic inscription "Kant" – so expressive in its brevity. He also remembered Jacobi, from whom the mathematical tradition of Königsberg derived as in Göttingen it derived from Gauss. He wanted a topic which would weave together these great names and all the separate strands of his career – Königsberg and Göttingen, Jacobi, Gauss, Kant, mathematics and science, science and experience, the great developments in knowledge and in thought through which he had lived.

Naturerkennen – the understanding of nature – *und Logik*. This would be his subject.

During the past decade he had become increasingly interested in reaching a greater audience with mathematical ideas. He had frequently accepted the opportunity of giving popular lectures in the Saturday morning series "for all the Faculties of the University." He took subjects like "Relativity Theory" or "The Infinite" or "The Principles of Mathematics" and tried, by finding examples from familiar fields outside mathematics, to make the fundamental concepts comprehensible to laymen.

"An enormous amount of labor was devoted to this task," Nordheim recalls. "We had to prepare preliminary outlines either of new material or from old lectures. These were then worked and re-worked practically every morning and in this process flavored with Hilbert's own inimitable brand of logic and humor."

During this period Hilbert and some of the other mathematics professors regularly attended the lectures of a zoologist. Hilbert had developed a great interest in genetics. He delighted in the laws determining the heredity of Drosophila, which could be obtained by the application of certain of the geometric axioms. He was fascinated by the Pferdespulwurm — "that creature with the most modest number of chromosomes, which corresponds therefore to the hydrogen atom with only one electron." But he was also impressed by the ability of the biologists to make their subject interesting and understandable to laymen.

"The biologists understand popular presentation especially well," he told Paul Funk one time. "In order to prevent the fatigue which straining thoughts bring forth in laymen, one must occasionally insert a little *dessin* (a French word meaning pattern, or design), and at that the biologists are pre-eminent." Pronouncing the French word sharply in his Königsberg dialect, he went on to say: "For us mathematicians, popular presentation is much harder, but still it must happen — if we go about it right — that we find a beautiful *dessin*."

Now, in the summer of 1930, among his mature fruit trees, he sought such a beautiful *dessin* as he began to prepare his speech for the Königsberg meeting, stripping away all vague generalities from his subject, putting his ideas into simple non-technical language for a general audience (that "man in the street" he had mentioned in Paris, to whom one should always be able to explain any fully realized mathematical theory).

He had returned many times over the years to his native city, but there was a special quality to this return. Kurt Reidemeister and Gabor Szegö, the mathematics professors now at the University, noted how pleased he was at the social gathering arranged in connection with his speech, so exuberant "that his wife must always again call him back." But Königsberg seemed colder than in the old days, and Hilbert had to borrow a fur coat from Szegö to keep warm.

The honorary citizenship was presented at the opening session. Then Hilbert took his place at the rostrum. His head was now almost entirely bald, the broad scholar's forehead contrasting more sharply than ever with the delicate chin; the white moustache and small beard were neatly trimmed. (Ostrowski was reminded of the head of Lenin.) He looked out at his audience through the familiar rimless glasses, the blue eyes still sharp and searching, and still so innocent. He placed his hands firmly on the manuscript in front of him and began to speak very slowly and carefully:

"The understanding of nature and life is our noblest task."

In recent times richer and deeper knowledge had been obtained in decades than had previously been obtained in the same number of centuries.

The science of logic had also progressed until there was now, in the axiomatic method, a general technique for the theoretical treatment of all scientific questions. Because of these developments — he told his audience — the men of today were better equipped than the philosophers of old to answer an ancient philosophical question: "the part which is played in our understanding by Thinking on the one side and Experience on the other."

It was a worthy question with which to conclude a career; for, fundamentally, to answer it would be to ascertain by what means general understanding is achieved and in what sense "all the knowledge which we collect in our scientific activities is truth."

Certain parallels between nature and thought had always been recognized. The most striking of these was a pre-established harmony which seemed to be almost the embodiment and realization of mathematical thought, the most magnificent and wonderful example of which was Einstein's theory of relativity.

But it seemed to him that the long recognized accord between nature and thought, experiment and theory, could only be understood when one took into account the formal element and the mechanism linked with it which exists on both sides, in nature and in thought. The extension of the methods of modern science should lead to a system of natural laws which corresponded with reality in every respect. Then we should need only pure thought — abstract deduction — in order to gain all physical knowledge. But this was not, in his opinion, the complete answer: "For what is the origin of these laws? How do we obtain them? How do we know that they correspond with reality? The answer is that we can obtain these laws only through our own experience . . . . Whoever wants nevertheless to deny that universal laws are derived from experience must contend that there is still a third source of understanding . . . ."

Königsberg's great son, Immanuel Kant, had been the classical exponent of this point of view — the point of view which Hilbert 45 years ago had defended at his public promotion for the degree of doctor of philosophy. Now, before his talk, he had smilingly commented to a young relative that a lot of what Kant had said was "pure nonsense" — but that, of course, he could not say to the citizens of Königsberg.

Kant had stated that man possesses beyond logic and experience certain *a priori* knowledge of reality.

"I admit," Hilbert told his audience, "that even for the construction of special theoretical subjects certain *a priori* insights are necessary . . . . I even believe that mathematical knowledge depends ultimately on some kind of such intuitive insight . . . . Thus the most general basic thought of Kant's theory of knowledge retains its importance . . . . The *a priori* is nothing more or less than . . . the expression for certain indispensable preliminary conditions of thinking and experiencing. But the line between that which we possess *a priori* and that for which experience is necessary must be drawn differently by us than by Kant — Kant has greatly overestimated the role and the extent of the *a priori*."

Men now knew that many facts previously considered as holding good *a priori* were not true, the most striking being the notion of an absolute present. But it had also been shown, through the work of Helmholtz and Gauss, that geometry was "nothing more than a branch of the total conceptual framework of physics." We had forgotten that the geometrical theorems were once experiences!

"We see now: Kant's *a priori* theory contains anthropomorphic dross from which it must be freed. After we remove that, only that *a priori* will remain which also is the foundation of pure mathematical knowledge."

In essence, this was the attitude which he had characterized in his recent work on the foundations of mathematics.

"The instrument which brings about the adjustment of differences between theory and practice, between thought and experiment, is mathematics. It builds the connecting bridge and continually strengthens it. Thus it happens that our entire present culture, insofar as it is concerned with the intellectual understanding and conquest of nature, rests upon mathematics!"

The effect of Hilbert's speech on the audience has been recalled by Oystein Ore, who, as a young man on his honeymoon, was there in Königsberg:

"I remember that there was a feeling of excitement and interest both in Hilbert's lecture and in the lecture of von Neumann on the foundations of set theory — a feeling that one now finally was coming to grips with both the axiomatic foundation of mathematics and with the reasons for the applications of mathematics in the natural sciences."

In the final part of his speech, Hilbert carefully made the point that in spite of the importance of the applications of mathematics, these must never be made the measure of its value. He concluded with that defense of *pure* mathematics which he had wanted so long ago to make in answer to the speech given by Poincaré at the first International Congress of Mathematicians.

"Pure number theory is that part of mathematics for which up to now no application has ever been found. But it is number theory which was considered by Gauss [who himself made untold contributions to applied mathematics] as the queen of mathematics . . . ."

Kronecker had compared the mathematicians who concerned themselves with number theory to Lotus-eaters who "once having consumed this food can never give it up."

"Even our great Königsberg mathematician Jacobi felt this way . . . . When the famous Fourier maintained that the purpose of mathematics lies in the explanation of natural phenomena, Jacobi objected, 'A philosopher like Fourier should know that the glory of the human spirit is the sole aim of all science!' . . . . Whoever perceives the truth of the generous thinking and philosophy which shines forth from Jacobi's words will not fall into regressive and barren scepticism."

Reidemeister and Szegö had made arrangements for Hilbert to repeat the conclusion of his speech over the local radio station; and when the session adjourned, they accompanied him to the broadcasting studio.

There, as Hilbert spoke into the unfamiliar instrument, it seemed that his voice rang out again with the enthusiasm and optimism of the vigorous man who in the prime of his life had sent his listeners out to seek the solution to 23 problems which, he was certain, would lead to progress in mathematics.

"In an effort to give an example of an unsolvable problem, the philosopher Comte once said that science would never succeed in ascertaining the secret of the chemical composition of the bodies of the universe. A few years later this problem was solved . . . .

"The true reason, according to my thinking, why Comte could not find an unsolvable problem lies in the fact that there is no such thing as an unsolvable problem."

He denied again, at the end of his career, the "foolish *ignorabimus*" of du Bois-Reymond and his followers. His last words into the microphone were firm and strong:

"Wir müssen wissen. Wir werden wissen."

We must know. We shall know.

As he raised his eyes from his paper and the technician snapped off the recording machine, he laughed.

The record which he made of this last part of his speech at Königsberg is still in existence. At the end, if one listens very carefully, he can hear Hilbert laugh.

"Wir müssen wissen. Wir werden wissen."

We must know. We shall know.

It was in every respect a great last line.

However, lives do not always end on great last lines.

At almost the same time that Hilbert was making his speech at Königsberg, a piece of work was being brought to a conclusion which was to deal a death blow to the specific epistemological objective of the final program of Hilbert's career. On November 17, 1930, the *Monatshefte für Mathematik und Physik* received for publication a paper by a 25-year-old mathematical logician named Kurt Gödel.

XXIII

Exodus

When Hilbert first learned about Gödel's work from Bernays, he was "somewhat angry."

The young man had taken up both of the problems of completeness which Hilbert had proposed at Bologna. He had established completeness for the case of the predicate calculus. But then he had proceeded to prove — with all the finality of which mathematics is uniquely capable — the *incompleteness* of the formalized number theory. He had also proved a theorem from which it follows that a finitist proof of consistency for a formal system strong enough to formalize all finitist reasonings is *impossible*.

In the highly ingenious work of Gödel, Hilbert saw, intellectually, that the goal toward which he had directed much effort since the beginning of the century — the final unanswerable answer to Kronecker and Brouwer and the others who would restrict the methods of mathematics — could not be achieved. Classical mathematics might be consistent and, in fact, probably was; but its consistency could never be established by mathematical proof, as he had hoped and believed it could be.

The boundless confidence in the power of human thought which had led him inexorably to this last great work of his career now made it almost impossible for him to accept Gödel's result emotionally. There was also perhaps the quite human rejection of the fact that Gödel's discovery was a verification of certain indications, the significance of which he himself had up to now refused to recognize, that the framework of formalism was not strong enough for the burden he wanted it to carry.

At first he was only angry and frustrated, but then he began to try to deal constructively with the problem. Bernays found himself impressed that even now, at the very end of his career, Hilbert was able to make great changes in his program. It was not yet clear just what influence Gödel's work would ultimately have. Gödel himself felt — and expressed the thought in his paper — that his work did not contradict Hilbert's formalistic point of view; and it soon became apparent that proof theory could still be fruitfully developed without keeping to the original program. Broadened methods would permit the loosening of the requirements of formalizing. Hilbert himself now took a step in this direction. This was the replacing of the schema of complete induction by a looser rule called "unendliche Induktion." In 1931 two papers in the new direction appeared.

Although he had retired, he continued to lecture regularly at the University. He still prepared in only the most general way, still frequently got stuck. When he found himself unable to work through a proof on the blackboard, he would dismiss it with a wave of his hand as "completely elementary." He sometimes stumbled over details, rambled impossibly, repeated himself. "But still, one out of three lectures was *superb*!"

Hilbert's career having come to its official end with his retirement, plans were made to begin the collecting and editing of his mathematical works. Blumenthal, who had observed and studied the personality and achievements of his teacher since 1895, was asked to compose a biography for the final volume. Although Blumenthal had been a professor at Aachen now for many years, he had never lost his strong feeling for Göttingen, returning time after time to be (as he said) "refreshed." Wherever he was, even at the front during the first world war, he always organized the former inhabitants of Göttingen into a social club. He took on the assignment of a life history with pleasure and painstaking care.

Volume I of the collected works was to be devoted to the *Zahlbericht* and the other number theory papers. For Hilbert, as for Gauss, the first years at Göttingen had been "the fortunate years." The papers on algebraic number fields were now recognized as the deepest and most beautiful of all his mathematical works. Helmut Hasse, who with Emmy Noether, van der Waerden, Artin, Takagi and others had taken part in carrying out the program for class-fields which Hilbert had outlined in the last number theory paper, was asked to write an evaluation of Hilbert's contribution in this area.

It seemed to Hasse, in retrospect, that Hilbert's work on algebraic number theory, like so much of his work, had stood in time and content at the turn of two centuries. On one side, treating problems in great generality with new methods which far surpassed the earlier methods in elegance and simplicity, he had thrown into relief the works of the number theorists of the old century. On the other side, "with wonderful farsightedness," he had sketched out paths to the positive final treatment of the whole complex of problems and had indicated the direction for the new century.

Three young mathematicians, trained in the Hilbert treatment of number theory by their teachers, were brought to Göttingen to assist with the editing. One of these was a young woman named Olga Taussky. Hilbert still enjoyed talking to young women. Mostly he talked to Fräulein Taussky about his health and about his wish to return some day and live out his life in Rauschen, the little Baltic fishing village where he had spent the vacations of his youth. But one day, looking back over his career and the many fields of mathematics in which he had worked, he remarked to her that, much as he admired all branches of mathematics, he considered number theory the most beautiful.

(That same year at the International Congress of Mathematicians at Zürich, in connection with a talk by his former student, Rudolf Fueter, Hilbert stated that the theory of complex multiplication of elliptic modular functions, which brings together number theory and analysis, was not only the most beautiful part of mathematics but also of all science.)

In the course of her work on Hilbert's papers, Fräulein Taussky was astonished to discover many errors. These were not typographical errors. Perhaps the bound of a function would be wrongly computed, a theorem incorrectly stated, a step omitted in a proof, or an entire proof necessary to the argument dismissed as "easily seen" when it was not. Although she recognized that, because of Hilbert's powerful mathematical intuition, the errors had not affected the ultimate results, she felt that they should be corrected in his collected works. She was encouraged in this by Emmy Noether, who was editing the work of Dedekind and frequently announced, loudly, that no one would be able to find a single error — "even with a magnifying glass!"

The number theory volume was to be presented to Hilbert on his seventieth birthday. A celebration was scheduled, a whole day of festivities. It was all very bothersome, Hilbert complained to Bernays, but it would be "good for mathematics."

Hermann Weyl wrote the birthday greeting that appeared in *Naturwissenschaften.* Throughout his scientific career, as he wrote to his old friend Robert König, he had kept before him a simple motto: "True to the spirit of Hilbert." The birthday of Hilbert, Weyl now noted in his greeting, was the high feast day for German mathematicians, celebrated year after year in warm personal veneration for the master but also in personal affirmation of their own beliefs and unity.

"Without doubt on the whole globe today Hilbert's name represents most concretely what mathematics means in the framework of objective spirits and how making mathematics, as a fundamental creative activity of mankind, is alive among us."

And yet, Weyl had to concede, Hilbert's own brand of optimism, his supreme confidence in the power of reason to come up with simple and clear answers to simple and clear questions, was "not popular nowadays" among the younger generation.

"Admittedly, one sentence or another of Hilbert's lecture [on logic and the understanding of nature, which he had given in Konigsberg in 1930] comes dangerously close to the opening words of Gottfried Keller's novel *Das Sinngedicht*, in which he mocks his scientist Reinhardt — 'About twenty-five years ago, when once again the natural sciences were at their highest peak . . . .'

"However, we do Hilbert an injustice if we confuse his rationalism with that of Haeckel . . . . He would rightly be called presumptious if, Faust-like, he had striven after the kind of magical knowledge which unlocks the very core of being to the intellect . . . . Such knowledge is different from the knowledge of reality that must prove itself by accurate prediction [and] can be advanced only by the mathematical method . . . .

"Hilbert seems to me to be an outstanding example of a man through whom the immensely creative power of naked scientific genius manifests itself . . . . I remember how enthralled I was by the first mathematics class I ever attended [at the University] . . . . It was Hilbert's famous course on the transcendence of e and π . . . .

"Woe to the youth that fails to be touched to the core by such a man as Hilbert!"

On the birthday itself, Ferdinand Springer, who was the publisher of the collected works, came to Göttingen to present personally to Hilbert the special white and gold leather-bound copy of the first volume. The beautiful cover contained, however, not the printed pages, but only the proofs of the pages; for Fräulein Taussky was still not satisfied. Hilbert made no comment on the unfinished nature of the volume. But later, in his presence, Fräulein Taussky declined a certain brand of cigarette as being too strong for her. Somebody said that one really couldn't tell one brand from another. "Aber nein!" Hilbert said. "Fräulein Taussky can tell the difference. She is capable of making the finest, the very finest distinctions." She was not sure, but she thought that he was making fun of her for taking so seriously errors which he himself considered unimportant.

On the birthday evening there was a party in the magnificent new building of the Mathematical Institute. Former colleagues and students came from all over Germany and many from abroad. Although it was during the Depression, everyone managed to look very elegant in shabby formal dress. Olga Taussky remembers that she purchased an evening gown secondhand for about two dollars, but it was much admired. There was a banquet with many loving speeches and toasts. Arnold Sommerfeld read to Hilbert a little verse which he had written: "Seiner Freunde treuester Freund / Hohler Phrase ärgster Feind." (*To his friends, truest friend / To the hollow phrase, bitterest enemy.*)

Then Hilbert made a short speech. He recalled the great good luck with which he had been blessed: the friendships with Minkowski and Hurwitz, the study time in Leipzig with Felix Klein, the Easter trip of 1888 when he had visited Gordan and Kronecker and many other mathematicians, his appointment by Althoff as Lindemann's successor at an unusually early age. And in his native city, he reminded the guests, he had had the good fortune to find his wife "who since then in faithful comradeship has taken a decisive part in my whole activity and especially in my concerns for the younger generation." Minkowski's name was mentioned frequently. His sudden death, Hilbert recalled, had left a "deep emptiness, both human and scientific," but life had had to go on. Edmund Landau had come to take Minkowski's place. Now Felix Klein's great goal had at last been achieved, and he himself was celebrating his seventieth birthday "in this beautiful Institute."

There was dancing after the banquet, and the guest of honor danced almost every dance. A procession of students carrying torches marched through the snow to the entrance of the brightly lighted building on Bunsen Strasse, and shouted for Hilbert. He came out and stood on the steps, bundled in his big coat with the fur collar, and somebody took a picture. From every window of the Institute famous faces looked out.

Here at the end was the highest honor which the students could give to a professor.

"For mathematics," Hilbert exhorted the shouting students, "hoch — hoch — hoch!" In English, it would have been "Hip hip hooray!"

A few days after the birthday celebration, Hasse expressed to Mrs. Hilbert "my ardent desire to talk once in my life personally to the great man." Mrs. Hilbert invited him to come to tea and afterwards left him alone in the garden with Hilbert.

"I began talking to him about what interested me most in those days — the theory of algebraic numbers and in particular class-field theory. On this theory I had written a report, in continuation of Hilbert's celebrated *Zahlbericht*; and I began telling him what I had done in this theory, based on his own famous results of the late nineties. But he interrupted me repeatedly and insisted that I explain to him the basic conceptions and results of that theory before he could listen to what I wanted to tell him. So I explained to him the very foundations of class-field theory. About this he got very enthusiastic and said, 'But that is extremely beautiful, who has created it?' And I had to tell him that it was he himself who had laid that foundation and envisaged that beautiful theory. After that he listened to what I had to tell him about my own results. He listened attentively, but more politely than intelligently."

In the Reichstag elections in the year of Hilbert's seventieth birthday, the National Socialist Party made great gains. The following January, President von Hindenburg appointed Adolf Hitler the chancellor of Germany. Almost immediately came the first measure designed to break that "satanical power" which had "grasped in its hands all key positions of scientific and intellectual as well as political and economic life." The universities were ordered to remove from their employment almost every full-blooded Jew who held any sort of teaching position.

The Hilbert school was perhaps the hardest hit. Hilbert's devotion to his science had always been complete. No prejudice — national, sexual or racial — had ever been allowed to enter into it. In 1917 an appropriate memorial had had to be written for Darboux even though his country was at war with Germany. A position had had to be obtained for Emmy Noether even though a woman had never been a Privatdozent at Göttingen. Since the earliest friendships with Minkowski and Hurwitz, scientists had never been classified by Hilbert as Aryan and non-Aryan. There were only two kinds of scientists: those who solved problems of recognized worth, and those who did not.

Now, in the Mathematical Institute itself, to whom did the ultimatum apply? To Courant, who had replaced Klein and brought to reality Klein's great dream. To Landau, who had come to Göttingen after the death of Minkowski and made the University the center of research in the theory of numbers. To Emmy Noether, who — in spite of the fact that she still received no more than a trifling stipend — was the center of the most fertile circle of research at that time in Göttingen. To Bernays, who had been Hilbert's assistant and collaborator for almost sixteen years. In the Physics Institute, both Born and Franck were Jews. A distinction, however, was made between them by the new government. Franck, who had already received his Nobel Prize, was exempted from the order. Born, who would not receive his Nobel for some years, had to go. The ultimatum applied to many others; sometimes it seemed to everyone.

Hilbert was extremely upset when he heard that many of his friends were being put on "forced leave," as the current euphemism had it.

"Why don't you sue the government?" he demanded of Courant. "Go to the state court? It is *illegal* for such a thing to happen!"

It seemed to Courant that Hilbert was completely unable to understand that lawlessness had taken over. Since his birthday, it had been hard to get him to listen and to accept innovations at the Institute. But chiefly his difficulty seemed to be that he still believed the old system of justice prevailed. He continued to retain the deep Prussian faith in the law which had been inculcated in him by Judge Hilbert. It is exemplified by the story of how when Frederick the Great was disturbed by the sound of a peasant's mill and threatened to confiscate it, the peasant replied to the king with complete confidence, "No — *in Prussia there are still judges*!" Frederick, ashamed, had had the peasant's words inscribed across the portico of his summer palace, where they still stood in 1933.

There was at first no general agreement among those affected about what was to be done. How far would it all go? "If you knew the German people, you knew it would go all the way." Young Hans Lewy decided to leave Germany when Hitler was appointed chancellor. By the first of April he was already in Paris. Some people who did not have to go left in protest. Franck aligned himself with his fellow Jews. Others thought that something of the greatness of Göttingen could still be salvaged. Landau was allowed to stay on because he had been a professor under the Empire. Further exceptions would be made. Courant had been gassed and wounded in the stomach fighting for Germany; surely that made him a German. Letters were sent to the Minister about the case of Fräulein Noether. She held such a minor job, received so little for her services. "I don't think there was ever such a distinguished list of recommendations," Weyl later said. Hilbert's name was at the top. But all the distinguished names had no effect.

"The so-called Jews are so attached to Germany," Hilbert said plaintively, "but the rest of us would like to leave."

Otto Neugebauer, now an assistant professor, was placed at the head of the Mathematical Institute. He held the famous chair for exactly one day, refusing in a stormy session in the Rector's office to sign the required loyalty declaration. The position of the head of the Mathematical Institute passed to Weyl. Although his wife was part Jewish, he was one of those who thought that something might yet be salvaged. All during the bitter uncertain spring and summer of 1933 he worked, wrote letters, interviewed officials of the government. But nothing could be changed.

By late summer nearly everyone was gone. Weyl, vacationing with his family in Switzerland, still considered returning to Göttingen in the hope that somehow he could keep alive the great scientific tradition. In America, his many friends worried about him and wrote long letters, advising, urging, begging that he leave Germany before it was too late. Abraham Flexner offered him a position at the Institute for Advanced Study. Finally Einstein, who had already been at the newly created Institute for several years, prevailed upon the younger man to come and join him there.

In Göttingen, Hilbert was left almost alone. He kept Bernays on as his assistant at his own expense. *The Foundations of Mathematics*, which he and Bernays had written in collaboration, was almost ready for publication. He put away his general mathematical books and became progressively more distant. With Bernays's help, he saw Arnold Schmidt and Kurt Schütte through the doctorate. Schütte was the last of 69 mathematicians (40 of them during the years from 1900 to 1914) to receive their degrees from Hilbert. In actuality, however, all of Schütte's contacts were through Bernays. He saw Hilbert only once.

"When I was young," Hilbert said to young Franz Rellich, one of the few remaining members of the old circle, "I resolved never to repeat what I heard the old people say — how beautiful the old days were, how ugly the present. I would never say that when I was old. But, now, I must."

Sitting next to the Nazis' newly appointed minister of education at a banquet, he was asked, "And how is mathematics in Göttingen now that it has been freed of the Jewish influence?"

"Mathematics in Göttingen?" Hilbert replied. "There is really none any more."

XXIV

Age

In the center of the town the swastika flew above the Rathaus and cast its shadow on the little goose girl. The university bulletin and publications appeared again in the old long-unused German type, the first page of each one bearing the statement that it appeared under the sponsorship of Herr Goebbels.

A Nazi functionary became the head of the Mathematical Institute. During the winter semester 1933–34 Hilbert lectured for one hour a week on the foundations of geometry. After the end of the semester, he never again came to the Institute.

Landau continued to lecture; but when he announced a course in calculus, an unruly mob prevented his entering the lecture hall. "It is all right for you to teach advanced courses," he was told, "but these are beginners and we don't want them taught by a Jew." Siegel, now a professor at Frankfurt, attempted to get support for his old teacher from a group of professors who were safe in their positions. He was not successful.

After a while Landau too was gone from Göttingen. Unlike the others, he did not leave the country, being tied to his native land by the fact of his wealth and possessions. Hardy arranged for him to deliver a series of lectures in England: "It was quite pathetic to see his delight when he found himself again in front of a blackboard and his sorrow when his opportunity came to an end."

By the spring of 1934 the situation had become so bad for the Jews that Bernays felt he must leave Germany, and he returned to Zürich. The Mathematical Institute continued to pay the salary of Hilbert's remaining assistant, Arnold Schmidt, who worked with him in his home on problems of logic and foundations.

"There were brief failings of memory which might make strangers think he was not so sharp," Schmidt says, "but those who worked with him in this area knew differently."

Now Helmut Hasse was made the head of the Institute. This was a great improvement; for although Hasse had long been a convinced nationalist, he was a first-rate mathematician.

That summer Emmy Noether, for whom a place had been found in America at Bryn Mawr, returned to Göttingen. "Her heart knew no malice," Weyl later explained. "She did not believe in evil — indeed it never entered her mind that it could play a role among men." Things were not so clear then as they were later to seem, and she wished Hasse only success in his efforts to rebuild the great tradition of Göttingen after the exodus of the previous year. At the end of the summer she returned to Bryn Mawr. She was at the height of her powers, her imagination and her technique having reached the maximum point of perfect balance. In her hands "the axiomatic method, no longer merely a method for logical clarification and deepening of the foundations, [had become] a powerful weapon of concrete mathematical research." Already, with it, she and van der Waerden and others had laid the foundations of modern algebra.

At first both the Hilberts had spoken out in such a forthright way against the new regime that their friends remaining in Göttingen were frightened for their safety. But they did not trust many of the people who were left, nor the new people who came, and after a while they too fell silent.

"Well, Herr Geheimrat, how do you fare?" one of the now infrequent visitors inquired of Hilbert.

"I — well, I don't fare too well. It fares well only with the Jews," he replied in the old unexpected way. "The Jews know where to stand."

Von Hindenburg died in the summer of 1934, purportedly leaving a will which bequeathed the presidency of the Reich to Hitler, who would then be both president and chancellor. An election was scheduled for August with the alternatives, *yes* or *no*. The day before the election, the newspapers carried a proclamation announcing that Hitler had the support of German science. The list of signatures included the name of Hilbert. Whether Hilbert actually signed the proclamation is not known. Arnold Schmidt, who was at that time seeing him almost every day, was not aware of the existence of such a proclamation until he was shown a copy of the newspaper report of it more than thirty years later. Signing would have been contrary to everything Schmidt knew from personal experience that Hilbert believed. He had to concede, however, that "at that time it is possible that Hilbert would have signed anything to get rid of someone who was bothering him."

In 1935 the final volume of the collected works, which contained the life history written by Blumenthal, was published. Hilbert wrote a little note to his oldest student, commenting on this last great piece of luck — that he should have such a splendid interpreter of his life and work. Blumenthal placed the note in his own copy of Hilbert's collected works.

For his biographical article, Blumenthal had called up his memories of his teacher since that day when the "medium-sized, quick, unpretentiously dressed man, who did not look at all like a professor" had come to Göttingen in the spring of 1895 as the successor of Heinrich Weber. But in spite of the warmth and affection, the life history remained objective.

"For the analysis of a great mathematical talent," Blumenthal concluded, "one has to differentiate between the ability to create new concepts and the gift for sensing the depth of connections and simplifying fundamentals. Hilbert's greatness consists in his overpowering, deep-penetrating insight. All of his works contain examples from far-flung fields, the inner relatedness of which and the connection with the problem at hand only he had been able to discern; from all these the synthesis — and his work of art — was ultimately created. As far as the creation of new things is concerned, I would place Minkowski higher, and from the classical great ones, for instance, Gauss, Galois, Riemann. But in his sense for discovering the synthesis only a very few of the great have equaled Hilbert."

In the spring of 1935 Emmy Noether died in the United States following an operation.

In his office at the Institute for Advanced Study, Einstein wrote a letter to the editor of The New York Times, which had reported her death only briefly: "In the judgment of most competent living mathematicians, Fräulein Noether was the most significant creative mathematical genius [of the female sex] thus far produced . . . .

"Beneath the effort directed toward the accumulation of worldly goods lies all too frequently the illusion that this is the most substantial and desirable end to be achieved; but there is, fortunately, a minority composed of those who recognize early in their lives that the most beautiful and satisfying experiences open to human kind are not derived from the outside but are bound up with the individual's own feeling, thinking and acting . . . . However inconspicuously the lives of these individuals run their course, nonetheless, the fruits of their endeavors are the most valuable contributions which one generation can make to its successors."

The editors of the *Annalen* decided to risk publishing a memorial article by van der Waerden. After the journal appeared in print, they waited for the blow to fall; but nothing happened. Taking courage, they published a paper by Blumenthal, who was still listed on the cover of the *Annalen* as one of the editors, although as a consequence of the Nuremberg Laws he had recently been removed from his professorship in Aachen. Again, nothing happened.

But the general scientific situation in Germany was progressively deteriorating. A strong supporter of the Third Reich was Ludwig Bieberbach, who had so passionately opposed the German mathematicians' attending the International Congress at Bologna. He and others analyzed the differences in the creative styles of German mathematicians and Jewish mathematicians. Death was no protection. When Klein was listed in a Jewish encyclopedia, his antecedents were carefully examined and it was determined finally that he was "a great German mathematician." Hilbert's antecedents were also examined. There was a joke that there was only one Aryan mathematician in Göttingen and in his veins Jewish blood flowed. The joke depended upon the fact that during Hilbert's illness, he had received a blood transfusion from Courant. Now the question was seriously raised if it was not suspicious for an Aryan mathematician to have the name *David*. It finally became necessary for Hilbert to produce the autobiography of Christian David Hilbert to show that David was a family name and that the other family names indicated that the Hilberts had at one time been Pietists.

In the late summer of 1936 the mathematicians of the world met again for another International Congress, this time at Oslo. Although Hilbert did not attend, he was remembered with a telegram of greeting from the delegates. Courant, who came from the United States, where he was now teaching at New York University, telephoned from Oslo; but Hilbert did not know what to say to his former pupil and colleague, and the conversation consisted of his fumbling over and over, "Well, what should I say? What should I ask now? Let me think for a moment."

In 1937 Hilbert was 75. A newspaper reporter came to interview him and ask him about places in Göttingen connected with the history of mathematics. "I actually know none," he said without (to the reporter's surprise) a trace of embarrassment at his ignorance. "Memory only confuses thought — I have completely abolished it for a long time. I really don't need to know anything, for there are others, my wife and our maid — they will know." As the reporter then began to express "a courteous doubt" whether one could so eliminate memory and history, Hilbert put back his head and gave a little laugh.

"Ja, probably I have even been known to be especially gifted for forgetting. For that reason indeed did I study mathematics."

Then he closed his eyes.

The reporter refrained from disturbing any further the old man, "the honorary doctor of five universities, who with easy serenity could completely forget everything — house, streets, city, names, occurrences and facts — because he had the power in each remaining moment to derive and develop again a whole world."

That night there was a birthday party at the Hilberts', a comparatively large affair for the new days. While the congratulatory speeches were being made, Hilbert sat in another room with his arms around the two young nurses who came regularly to the house to give him some physical treatments. When Hecke, who had come from Hamburg for the party, reminded him that he really should listen to the glowing speeches being made about him and his work, he laughed, "This is much better!"

Elizabeth Reidemeister took a birthday picture. She reminded him of what seemed to her some important event in which they had both participated, and was surprised that he remembered nothing of it. "I am interested only in the stars," he explained.

During this period Franz was home again. With age he had come to look disconcertingly like his father. He patterned himself after him, loudly spoke out his opinion on all subjects — a tragic parody — "the sound without the substance," as the people in Göttingen observed. He never held any real job.

But he also studied various subjects very thoroughly — Goethe, theology. He was a real "Kenner," according to Arnold Schmidt — an expert on the fields of his interests. He spoke often of learning mathematics so that he could appreciate his father's work.

The next year — 1938 — saw the last birthday party in the house on Wilhelm Weber Strasse. There were only a few old friends for lunch. Hecke came from Hamburg, Carathéodory from Munich. Siegel, who was now at the Institute in Göttingen, was there. Also present was Blumenthal.

"What subjects are you lecturing on this semester?" Hilbert asked.

"I do not lecture any more," Blumenthal gently reminded him.

"What do you mean, you do not lecture?"

"I am not allowed to lecture any more."

"But that is completely impossible! That cannot be done. Nobody has a right to dismiss a professor unless he has committed a crime. Why do you not apply for justice?"

The others tried to explain Blumenthal's situation, but Hilbert became increasingly angry with them.

"I felt," Siegel says, "that he had the impression we were trying to play a bad joke on him."

Shortly afterwards, Blumenthal's name had to be removed from the cover of the *Annalen*. With the help of his friends he left Germany for Holland.

Across the ocean in America, George Pólya, who was now at Stanford University, reminded Weyl that it was 1938 and that by its terms their wager on the future of Intuitionism was now up. Weyl conceded that he had indeed lost, but he asked Pólya please not to make him concede publicly.

That same year Edmund Landau died.

Life went on in Göttingen. Mrs. Hilbert, who was gradually losing her sight, mourned that the people who used to come to visit no longer came. Sometimes, though, there were still small social gatherings around the Hilberts.

On one of these occasions there was a discussion about which German town was the most beautiful. Some of the guests said Dresden, others said Munich. But Hilbert insisted, "No, no — the most beautiful town in Germany is still Königsberg!" When his wife protested, "But, David, you can't really say that — Königsberg is not all that beautiful," he replied: "But, Käthe, after all I must know, for I have spent my whole life there." Even when she reminded him that in fact they had come to Göttingen more than forty years before, he shook his head: "Ah, a few little years — I have spent my whole life in Königsberg!"

"So," Hasse, who was present, thought sadly, "his mind has condensed all the forty fruitful years of his wonderful achievement in so many branches of mathematics into *a few little years*."

In 1939 an agreement signed in Munich among Germany, France, England and Italy seemed for the moment to guarantee "peace in our time." The Swedish Academy of Science announced the first Mittag-Leffler Prize, which was to go to David Hilbert and Émile Picard. The old Frenchman received his prize from the emissary of the Academy, Torsten Carleman, at a large banquet in Paris. After a glowing tribute, Carleman presented Picard with a complete, beautifully bound set of Mittag-Leffler's *Acta*

Mathematica, the journal which had been one of the first to publish Cantor's works. From Paris, Carleman went to Göttingen. He expected that there he would enjoy a repetition of banquet, speech and presentation. Disappointed when such was not forthcoming, he still insisted that he wanted to award the prize to Hilbert in person. Hasse and Siegel finally located Hilbert, who was in the nearby Harz mountains with his wife, and drove Carleman to the inn where they were staying. Hilbert listened silently to Carleman's tribute. Shortly afterwards, the 72 red leather volumes of *Acta Mathematica* appeared on the shelves of the library of another mathematician, to whom Hilbert had almost immediately sold them.

It was August again. On the first day of September Germany invaded Poland. Within a week France and England had declared war on Germany.

Hilbert's assistant now was the gifted young logician, Gerhard Gentzen, who, following the new and less restrictive methods of "transfinite induction," had been able to achieve the long-sought proof of the consistency of arithmetic. The proof had been managed, however, only by substantially lowering the standards Hilbert had originally set up. Gentzen came regularly to Hilbert's house and read aloud – at Hilbert's request – the poems of Schiller. After a while Gentzen too was gone. He died in 1945 following arrest and imprisonment in Prague.

Holland was invaded. In England efforts were made by Ewald and others to get Blumenthal to safety. But it was too late.

Siegel had vowed at the end of the first war that he would not remain in Germany during another war. In March 1940, he received an invitation to deliver a lecture in Oslo. He knew that he would not see the Hilberts again, and so he went to say goodbye. They were not at the house on Wilhelm Weber Strasse, the furnace having broken down; but he found them in a shabby hotel where, Mrs. Hilbert told him, Hermann Amandus Schwarz always used to stay when he was in Göttingen. Schwarz had been responsible for bringing Klein to the University and had been dead now some twenty years. The Hilberts were having breakfast in their room, Hilbert sitting on the bed and eating from a jar of caviar that Niels Bohr had sent him from Copenhagen. Siegel said goodbye. In Oslo he found that the Bohr brothers and Oswald Veblen had already arranged passage to the United States, where a place at the Institute for Advanced Study would be waiting for him. Two days after he left Oslo, the Germans invaded Norway.

In December 1941, a month before Hilbert's eightieth birthday, the United States entered the war. Although there was no party on the eightieth birthday, a tribute to Hilbert appeared as usual. It was prepared by Walther Lietzmann, who in 1902 had headed the delegation from the Mathematics Club who had pleaded with Hilbert to refuse the tempting offer from Berlin and remain in Göttingen. The story of Hilbert's life and his achievements, everything was there in Lietzmann's tribute except the names of the many Jews (other than Minkowski and Hurwitz) who had played such an important part in his career. Blumenthal was circumspectly quoted merely as "the author of the life history in the collected works." The picture that accompanied the tribute was a recent one, and the eyes which had looked so firmly and innocently out at the world seemed now distrustful.

In Holland, Blumenthal dedicated a paper to Hilbert in honor of the eightieth birthday.

The Berlin Academy voted to commemorate the birthday with a special citation for that work which of all the influential Hilbert works had had the most pervasive influence on the progress of mathematics — the little 92-page book on the foundations of geometry.

The day that this award was voted by the Academy, Hilbert fell on the street in Göttingen and broke his arm. He died, a little more than a year later, on February 14, 1943, of complications arising from the physical inactivity that resulted from the accident.

Not more than a dozen people attended the morning funeral service in the living room of the house on Wilhelm Weber Strasse. From Munich came one of his oldest friends. Standing beside the coffin, Arnold Sommerfeld spoke of Hilbert's work.

What had been his greatest mathematical achievement?

"The invariants? The number theory, which was so loved by him? The axiomatics of geometry, which was the first great achievement in this field since Euclid and the non-euclidean geometries? What Riemann and Dirichlet surmised, Hilbert established by proof at the foundation of function theory and the calculus of variations. Or were integral equations the high point . . . . Soon in the new physics . . . they bore most beautiful fruits. His gas theory had a fundamental effect on the new experimental knowledge, which has not yet been played out. Also his contributions to general relativity theory are of permanent value. Of his final endeavors in connection with mathematical knowledge, the last word has still not been spoken. But when in this field a further development is possible, it will not by-pass Hilbert but go through him."

Carathéodory had also planned to come from Munich for the funeral, but he had fallen ill. The tribute which he had written was read by Gustav Herglotz with tears streaming down his face.

Above all, it had seemed to Carathéodory, there had been such an absolute integrity and oneness in the conduct of Hilbert's life that "even the peculiarities of his old age, which might cause a stranger surprise, were for us, his friends, genuine manifestations of the Hilbert character."

The active influence which Hilbert had exercised on the mathematicians of his time was embodied for Carathéodory in a statement he had heard one of the most important mathematicians of the day make directly to Hilbert himself: "You have made us all think only that which you would have us think!"

Recalling what the dead man had meant to them during his lifetime, he addressed the widow directly:

"Dear Frau Hilbert, now he lies in this same room in which we spent so many joyous hours with him. In a short time we will carry him to his last resting place. But so long as our hearts beat, we shall be bound together by the memory of this great man."

They buried Hilbert in the cemetery out beyond the river, where Klein also lay. Only the name and dates were inscribed on a plaque in the grass.

The news of Hilbert's death came to the outside warring world by way of Switzerland. Attending a mathematical meeting in New York, Hermann Weyl saw a small notice from Bern in The Times. He recalled again the summer months which he had spent long ago working his way through the *Zahlbericht*, the happiest months of his life, "whose shine, across years burdened with our common share of doubt and failure, still comforts my soul." Back home in Princeton, he wrote to Auguste Minkowski, who was living in Boston with her older daughter:

"The report of Hilbert's death brought up again the whole Göttingen past for me. I had the great luck to grow up in the most beautiful years . . . when Hilbert and your husband both stood in the prime of their power . . . . I believe it has very seldom happened in mathematics that two men exercised such a strong and magic influence on a whole generation of students. It was a beautiful, brief time. Today there is nowhere anything even remotely comparable . . . ."

The story of Hilbert's death came to bombed England more slowly and less accurately.

"The news has reached us that David Hilbert died during the summer (sic)," the *Journal of the London Mathematical Society* announced that fall. "The Council feels that the death of so great a mathematician should not pass, even momentarily, unnoticed; but the difficulties of obtaining an adequate account of his work and influence are at present unsurmountable."

Max Born, who was in England, had been in Göttingen when Minkowski died; and now he could not help wondering: "Hilbert had survived his friend by more than thirty years. He was still permitted to achieve important work. But who would like to say if his solitary death in the dark Nazi time was not still more tragic than Minkowski's in the fullness of his power?"

A few months after Hilbert's death, Blumenthal was swept up by the Gestapo in one of its periodic raids on the Jews in Holland. He was sent to Theresienstadt, a little Czechoslovakian village which had been converted into a ghetto for old Jews and others for whose deaths the Nazis – at least in the beginning – did not wish to take direct responsibility. It is known that at one time he was placed on a train for Auschwitz, then for some reason removed before the train left. He died in Theresienstadt at the end of 1944.

On January 17, 1945, Käthe Hilbert died. She was almost blind. There was no old friend to speak by her coffin; and since she, like her husband, had long ago left the church, no minister would perform this office. In the end, at the begging of Franz Hilbert, a woman who had never known the great days spoke a few words.

That same year Königsberg, almost completely destroyed, fell to the Russians.

XXV

The Last Word

It might seem that the sweet sound of the Pied Piper had been stilled forever. But all over the world — in little European countries — in embattled England — Japan — Russia — the United States — there were Hilbert students, and students of Hilbert students.

Across the ocean, even during the war, one could still hear the old tune. Hermann Weyl, with his own special ardor, was attempting to create in Princeton at the Institute for Advanced Study another great center of passionate scientific life — it was his phrase — like the one he had known in his youth in Göttingen. In New York, Richard Courant was ensconced in an abandonned hat factory, called cheerfully by his friends "Courant's Institute." The spirit of Hilbert lived there too.

At the time of Hilbert's death it was said in *Nature* that there was scarcely a mathematician in the world whose work did not derive in some way from that of Hilbert. Like some mathematical Alexander, he had left his name written large across the map of mathematics. There was, as *Nature* pointed out, Hilbert space, Hilbert inequality, Hilbert transform, Hilbert invariant integral, Hilbert irreducibility theorem, Hilbert base theorem, Hilbert axiom, Hilbert sub-groups, Hilbert class-field.

The ideas in his work on Gordan's Problem had extended far beyond algebraic invariants in method and significance: they had flowered into the general theories of abstract fields, rings and modules — in short, modern algebra. Much of the work of subsequent number theorists had been in the fertile fields which Hilbert had opened up with the *Zahlbericht* and his program of class-fields. The little book on the foundations of geometry — "a landmark in mathematical thought" — had set the axiomatic method securely and deeply in nearly all of mathematics. "It is difficult," a recent mathematical historian has commented, "to overestimate the influence of this little book." Since Hilbert's salvaging of the Dirichlet Principle, the theory had been simplified and extended until the Principle had become "a tool as flexible and almost as simple as that originally envisaged by Riemann." The work had also been the starting point for the development of what are known as direct methods in the variational calculus, important in both pure and applied mathematics. The theorem which Hilbert had stated and proved in his lectures on the calculus of variations at the turn of the century had been used to lay a whole new foundation for that subject. The general theory of integral equations had become one of the most powerful mathematical weapons in the arsenal of the physicists, and Hilbert Space theory had grown to such proportions that a writer complained it was impossible to treat it "in a finite number of words." And although, today, no one speaks of the axiomatization of physics, the clarifying, ordering and unifying power of the axiomatic method has penetrated that science.

But what of the last great work of Hilbert's career of formalizing mathematics and establishing its consistency by absolute proof? In spite of the blow dealt to this program by Gödel's work, Hilbert's liberating conception of mathematical existence as freedom from contradiction has unquestionably triumphed over the shackling constructive ideas of his opponents. The question of the consistency of mathematics, so seemingly simple and obvious until it was raised by Hilbert, has played a role of inestimable importance in the history of mathematical thought. "It was a good question," one modern mathematician says, "and only a very great mathematician would have thought of asking it."

Gödel (who never met nor had any correspondence with Hilbert) feels that Hilbert's scheme for the foundations of mathematics "remains highly interesting and important in spite of my negative results."

He adds:

"What has been proved is only that the *specific epistemological* objective which Hilbert had in mind cannot be obtained. This objective was to prove the consistency of the axioms of classical mathematics on the basis of evidence just as concrete and immediately convincing as elementary arithmetic.

"However, viewing the situation from a purely *mathematical* point of view, consistency proofs on the basis of suitably chosen stronger metamathematical presuppositions (as have been given by Gentzen and others) are just as interesting, and they lead to highly important insights into the proof theoretic structure of mathematics. Moreover, the question remains open whether, or to what extent, it is possible, on the basis of the formalistic approach, to prove 'constructively' the consistency of classical mathematics, i.e., to replace its axioms about abstract entities of an objective Platonic realm by insights about the given operations of our mind.

"As far as my negative results are concerned, apart from the philosophical consequences mentioned before, I would see their importance primarily in the fact that in many cases they make it possible to judge, or to guess, whether some specific part of Hilbert's program can be carried through on the basis of given metamathematical presuppositions."

Gödel also feels that "in judging the value of Hilbert's work on the Continuum Problem, it is frequently overlooked that, disregarding questions of detail, one quite important *general* idea of his has proved perfectly correct, namely that the Continuum Problem will require for its solution entirely new methods deriving from the foundations of mathematics. This, in particular, would seem to imply (although Hilbert did not say so explicitly) that the Continuum Hypothesis is undecidable from the usual axioms of set theory."

As a result of Hilbert's enthusiasm for the problems of mathematical logic and the foundations of mathematics, a whole new area of study has been added to the science – metamathematics, *beyond mathematics*.

"The future historian of science concerned with the development of mathematics in the late nineteenth and the first half of the twentieth century will undoubtedly state that several branches of mathematics are highly indebted to Hilbert's achievements for their vigorous advancement in that period," Alfred Tarski has written. "On the other hand, he will have to note, perhaps with some wonder, that the influence of this man appears equally strong and powerful in some other domains which do not owe any exceptionally important results to Hilbert's own research. An example of this kind is furnished by the foundations of geometry. I am far from underestimating the value of Hilbert's contributions . . . in his [*Foundations of Geometry*], but I think that his most essential merit was the impulse he gave to organized research in this domain. A still more striking example is presented by metamathematics. Occasional considerations in this field preceded Hilbert's Paris address; the first positive and really profound results appeared before Hilbert started his continuous work in this domain... [and] one does not immediately associate with Hilbert's name any definite and important metamathematical result. Nevertheless, Hilbert will deservedly be called the father of metamathematics. For he is the one who created metamathematics as an independent being; he fought for its right to existence, backing it with his whole authority as a great mathematician. And he was the one who mapped out its future course and entrusted it with ambitions and important tasks. It is true that the baby did not fulfill all the expectations of the father, it did not grow up to be a child prodigy. But it developed sanely and healthily, it has become a normal member of the great mathematical family, and I do not think that the father has any reason to blush for his progeny . . . ."

In 1950, when Hermann Weyl was asked by the American Mathematical Society to summarize the history of mathematics during the first half of the twentieth century, he wrote that if the terminology of the Paris Problems had not been so technical he could have performed the required task simply in the terms of Hilbert's problems which had been solved, or partially solved — "a chart by which we mathematicians have often measured our progress" during the past fifty years. "How much better he predicted the future of mathematics than any politician foresaw the gifts of war and terror that the new century was about to lavish on mankind!"

Today, Königsberg no longer exists. Where it stood on the Pregel there now stands Kaliningrad, the most advanced naval base of the Soviet Union. The rivalry between Göttingen and Paris is a thing of the past. In both Germany and France a whole generation of mathematicians is missing. The United States finds itself immeasurably enriched, for almost all of the members of the Hilbert school and many other European scientists emigrated to that country. Among them were the following who have been mentioned at times in the course of this book: Artin, Courant, Debye, Dehn, Einstein, Ewald, Feller, Franck, Friedrichs, Gödel, Hellinger, von Kármán, Landé, Lewy, Neugebauer, von Neumann, Emmy Noether, Nordheim, Ore, Pólya, Szegö, Tarski, Olga Taussky, Weyl, Wigner.

After the war Göttingen was the first German university to re-open its doors. Eventually many of Hilbert's old friends returned to visit there, a few like Max Born to live out their lives nearby.

In 1962, on the one hundredth anniversary of Hilbert's birth, Richard Courant delivered a talk in Göttingen on Hilbert's work and its importance for mathematics.

"It is naturally impossible in a program like this to give an even approximate appreciation of a personality so many-sided as that of Hilbert," he said. "Also there is no point in attempting sentimentally to bring back the good old times. But I feel that the consciousness of Hilbert's spirit is of great actual importance for mathematics and mathematicians today.

"Although mathematics has played an important role for more than two thousand years, it is still subject to changes of fashion and, above all, to departures from tradition. In the present era of the over-active industrialization of science, propaganda, and the explosive manipulation of the social and personal basis of science, I believe that we find ourselves in such a period of danger. In our time of mass media, the call for reform, as a result of propaganda, can just as easily lead to a narrowing and choking as to a liberating of mathematical knowledge. That applies, not only to research in the universities, but also to instruction in the schools. The danger is that the combined forces so press in the direction of abstraction that only that side of the great Hilbertian tradition is carried on.

"Living mathematics rests on the fluctuation between the antithetical powers of intuition and logic, the individuality of 'grounded' problems and the generality of far-reaching abstractions. We ourselves must prevent the development being forced to only one pole of the life-giving antithesis.

"Mathematics must be cherished and strengthened as a unified, vital branch in the broad river of science; it dares not trickle away in the sand.

"Hilbert has shown us through his impressive example that such dangers are easily preventable, that there is no gap between pure and applied mathematics, and that between mathematics and science as a whole a fruitful community can be established. I am therefore convinced that Hilbert's contagious optimism even today retains its vitality for mathematics, which will succeed only through the spirit of Hilbert."

It is this optimism which will echo, as long as stone survives, from the marker that has been placed since Hilbert's death over his grave in Göttingen:

Wir müssen wissen.
Wir werden wissen.

Otto Hilbert, father of David Hilbert, as a university student in 1850

The Königsberg cathedral (reprinted from *Ostpreussen, Westpreussen, Danzig*, Gräfe und Unzer Verlag, München)

The Pregel river with the Königsberg castle in the background (reprinted from *Ostpreussen, Westpreussen, Danzig*, Gräfe und Unzer Verlag, München)

Hermann Minkowski when he won the prize of the Paris Academy

Adolf Hurwitz as an Extraordinarius in Königsberg

David Hilbert, 1886

David Hilbert and
Käthe Jerosch, 1892

David Hilbert, c. 1900

Franz Hilbert, only son of
David and Käthe Hilbert

The Mathematics Club of Göttingen, 1902
Left to right, front row: Abraham, Schilling, Hilbert, Klein, Schwarzschild, Mrs. Young, Diestel, Zermelo; second row: Fanla, Hansen, C. Müller, Dawney, E. Schmidt, Yoshiye, Epsteen, Fleisher, F. Bernstein; third row: Blumenthal, Hamel, H. Müller

Felix Klein during his Leipzig days

Hermann Minkowski

David Hilbert, 1912 — one of a group of portraits of professors which were sold as postcards in Göttingen

Edmund Landau

A dinner party at the Kleins' with Paul Gordan (far left), Klein (center) and Käthe Hilbert (far right)

Richard Courant as a student in Göttingen

Carl Runge

Max Born as a Privatdozent in Göttingen

Emmy Noether

Constantin Carathéodory

Hilbert's sixtieth birthday party
Front row, left to right: Richard Courant, Franz Hilbert, Mrs. Courant (Nina Runge), Hertha Sponer (later Mrs. Franck), Mrs. Grotrian; second row: Mrs. Esslen (later Mrs. Springer), Mrs. Landau, Mrs. Hilbert, David Hilbert, Mrs. Hofmann, Mrs. Minkowski; third row: Ferdinand Springer, Felix Bernstein (behind Mrs. Landau), Mrs. Prandtl, Edmund Landau, Mrs. Franck, Fanny Minkowski (at end of row); fourth row: Ernst Hellinger, Erich Hecke (behind Landau), Walter Grotrian (behind Mrs. Hofmann); fifth row: Peter Debye, Theodore von Kármán (behind Hecke), Mrs. Rüdenberg (Lily Minkowski), Paul Bernays, Leonard Nelson, "Klärchen" (second from end of row).

David Hilbert and Hermann Weyl during the mid-twenties

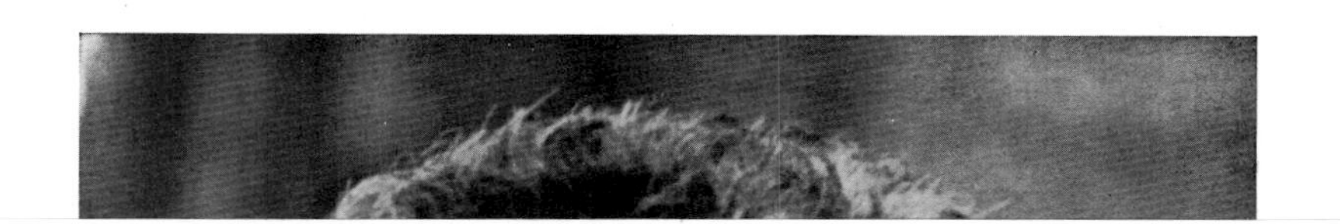

The entrance to the Mathematical Institute of Göttingen

Rear view of the Mathematical Institute

Hermann Weyl

David Hilbert, 1932

David Hilbert and His Mathematical Work [1]

By Hermann Weyl

Bulletin of the American Mathematical Society 50, 612—654 (1944)
Boletin da Sociedade de Matemática de São Paulo
1, 76—104 (1946) and 2, 37—60 (1947)

A great master of mathematics passed away when David Hilbert died in Göttingen on February the 14th, 1943, at the age of eighty-one. In retrospect it seems to us that the era of mathematics upon which he impressed the seal of his spirit and which is now sinking below the horizon achieved a more perfect balance than prevailed before and after, between the mastering of single concrete problems and the formation of general abstract concepts. Hilbert's own work contributed not a little to bringing about this happy equilibrium, and the direction in which we have since proceeded can in many instances be traced back to his impulses. No mathematician of equal stature has risen from our generation.

America owes him much. Many young mathematicians from this country, who later played a considerable role in the development of American mathematics, migrated to Göttingen between 1900 and 1914 to study under Hilbert. But the influence of his problems, his viewpoints, his methods spread far beyond the circle of those who were directly inspired by his teaching.

Hilbert helped the reviewer of his work greatly by seeing to it that it is rather neatly cut into different periods during each of which he was almost exclusively occupied with one particular set of problems. If he was engrossed in integral equations, integral equations seemed everything; dropping a subject, he dropped it for good and turned to something else. It was in this characteristic way that he achieved universality. I discern five main periods: i. Theory of invariants (1885—1893). ii. Theory of algebraic number fields (1893—1898). iii. Foundations, (a) of geometry (1898—1902), (b) of mathe-

[1] This contribution is reproduced in a shortened version.

matics in general (1922–1930). iv. Integral equations (1902—1912). v. Physics (1910–1922). The headings are a little more specific than they ought to be. Not all of Hilbert's algebraic achievements are directly related to invariants. His papers on calculus of variations are here lumped together with those on integral equations. And of course there are some overlappings and a few stray children who break the rules of time, the most astonishing his proof of Waring's theorem in 1909.

His Paris address on "Mathematical problems" . . . straddles all fields of our science. Trying to unveil what the future would hold in store for us, he posed and discussed twenty-three unsolved problems which have indeed, as we can now state in retrospect, played an important role during the following forty odd years. A mathematician who had solved one of them thereby passed on to the honors class of the mathematical community.

Literature

Hilbert's *Gesammelte Abhandlungen* were published in 3 volumes by J. Springer, Berlin, 1932—35. This edition contains his *Zahlbericht*, but not his two books:

Grundlagen der Geometrie, 7th ed., Leipzig, 1930.

Grundzüge einer allgemeinen Theorie der linearen Integralgleichungen, Leipzig and Berlin, 1912.

Hilbert is co-author of the following works:

R. Courant and D. Hilbert, *Methoden der mathematischen Physik*, Berlin, vol. 1, 2d ed., 1931, vol. 2, 1937.

D. Hilbert and W. Ackermann, *Grundzüge der theoretischen Logik*, Berlin, 1928.

D. Hilbert and S. Cohn-Vossen, *Anschauliche Geometrie*, Berlin, 1932.

D. Hilbert and P. Bernays, *Grundlagen der Mathematik*, Berlin, vol. 1, 1934, vol. 2, 1939.

The Collected Papers contain articles by B. L. van der Waerden, H. Hasse, A. Schmidt, P. Bernays, and E. Hellinger on Hilbert's work in algebra, in number theory, on the foundations of geometry and arithmetics, and on integral equations. These articles trace the development after Hilbert, giving ample references. The reader may also consult a number of Die Naturwissenschaften vol. 10 (1922) pp. 65–104, dedicated to Hilbert, which surveys his work prior to 1922, and an article by L. Bieberbach, *Ueber den Einfluss von Hilberts Pariser Vortrag über "Mathematische Probleme" auf die Entwicklung der Mathematik in den letzten dreissig Jahren*, Die Naturwissenschaften vol. 18 (1930) pp. 1101–1111. O. Blumenthal wrote a life of Hilbert (Collected Papers, vol. 3, pp. 388—429).

I omit all quotations of literature covered by these articles.

The classical theory of invariants deals with polynomials $J = J(x_1 \cdots x_n)$ depending on the coefficients $x_1, \cdots, x_n$ of one or several ground forms of a given number of arguments $\eta_1, \cdots, \eta_g$. Any linear substitution s of determinant 1 of the g arguments induces a certain linear transformation $U(s)$ of the variable coefficients $x_1, \cdots, x_n$, $x \to x' = U(s)x$, whereby $J = J(x_1 \cdots x_n)$ changes into a new form $J(x'_1 \cdots x'_n) = J^e(x_1 \cdots x_n)$. J is an invariant if $J^e = J$ for every s. (The restriction to *unimodular* transformations s enables one to avoid the more involved concept of *relative* invariants and to remove the restriction to homogeneous polynomials, with the convenient consequence that the invariants form a *ring*.) The classical problem is a special case of the general problem of invariants in which s ranges over an arbitrary given abstract group Γ and $s \to U(s)$ is any representation of that group (that is, a law according to which every element s of Γ induces a linear transformation $U(s)$ of the n variables $x_1, \cdots, x_n$ such that the composition of group elements is reflected in composition of the induced transformations). The development before Hilbert had led up to two main theorems, which however had been proved in very special cases only. The first states that *the invariants have a finite integrity basis*, or that we can pick a finite number among them, say $i_1, \cdots, i_m$, such that every invariant J is expressible as a polynomial in $i_1, \cdots, i_m$. An identical relation between the basic invariants $i_1, \cdots, i_m$ is a polynomial $F(z_1 \cdots z_m)$ of m independent variables $z_1, \cdots, z_m$ which vanishes identically by virtue of the substitution

$$z_1 = i_1(x_1 \cdots x_n), \cdots, z_m = i_m(x_1 \cdots x_n).$$

The second main theorem asserts that *the relations have a finite ideal basis*, or that one can pick a finite number among them, say $F_1, \cdots, F_h$, such that every relation F is expressible in the form

$$F = Q_1 F_1 + \cdots + Q_h F_h, \tag{1}$$

the Q_i being polynomials of the variables $z_1, \cdots, z_m$.

I venture the guess that Hilbert first succeeded in proving the *second* theorem. The relations F form a subset within the ring $k[z_1 \cdots z_m]$ of all polynomials of $z_1, \cdots, z_m$ the coefficients of which lie in a given field k. When Hilbert found his simple proof he could not fail to notice that it applied to any set of polynomials Σ whatsoever and he thus discovered one of the most fundamental theorems of algebra, which was instrumental in ushering in our modern abstract approach, namely:

(A) *Every subset Σ of the polynomial ring $k[z_1 \cdots z_m]$ has a finite ideal basis.*

Is it bad metaphysics to add that his proof turned out so simple *because* the proposition holds in this generality? The proof proceeds by the adjoining of one variable z_i after the other, the individual step being taken care of by the statement: If a given ring r satisfies the condition (P): that every subset of r has a finite ideal basis, then the ring $r[z]$ of polynomials of a single variable z with coefficients in r satisfies the same condition (P). Once this is established one gets not only (A) but also an arithmetic refinement discussed by Hilbert in which the field k of rational numbers is replaced by the ring of rational integers.

The subset Σ of relations to which Hilbert applies his theorem (A) is itself an *ideal*, and thus the ideal $\{F_1, \cdots, F_h\}$, that is, the totality of all elements of the form (1), $Q_i \varepsilon k[z_1 \cdots z_m]$, not only contains, but coincides with, Σ. The proof, however, works even if Σ is not an ideal, and yields at one stroke (1) the enveloping ideal $\{\Sigma\}$ of Σ and (2) the reduction of that ideal to a finite basis, $\{\Sigma\} = \{F_1, \cdots, F_h\}$.

Construction of a full set of relations $F_1, \cdots, F_h$ would finish the investigation of the algebraic structure of the ring of invariants were it true that any relation F can be represented in the form (1) *in one way only*. But since, generally speaking, this is not so, we must ask for the "vectors of polynomials" $M = (M_1, \cdots, M_h)$ for which $M_1F_1 + \cdots + M_hF_h$ vanishes identically in z (syzygy of first order). These linear relations M between $F_1, \cdots, F_h$ again form an ideal to which Theorem (A) is applicable, the basis of the M thus obtained giving rise to syzygies of the second order. To the first two main theorems Hilbert adds a third to the effect that if redundance is avoided, the chain of syzygies breaks off after at most m steps.

All this hangs in the air unless we can establish the *first main theorem*, which is of an altogether different character because it asks for an integrity, not an ideal basis. Discussing invariants we operate in the ring $k_x = k[x_1 \cdots x_n]$ of polynomials of $x_1, \cdots, x_n$ in a given field k. Hilbert applies his Theorem (A) to the totality $\mathfrak{J}$ of all *invariants J for which $J(0, \cdots, 0) = 0$* (a subring of k_x which, by the way, is not an ideal!) and thus determines an ideal basis $i_1, \cdots, i_m$ of $\mathfrak{J}$. Each of the invariants $i = i_r$ may be decomposed into a sum $i = i^{(1)} + i^{(2)} + \cdots$ of homogeneous forms of degree 1, 2, $\cdots$, and as the summands are themselves invariants, the i_r may be assumed to be homogeneous forms of degrees $\nu_r \geqq 1$. Hilbert then claims that the $i_1, \cdots, i_m$ constitute an integrity basis for all invariants. I use a finite group Γ consisting of N elements s (although this case of the general problem of invariants was never envisaged by Hilbert himself) in order to illustrate the idea by which the transition is made. Every invariant J is representable in the form

$$J = c + L_1 i_1 + \cdots + L_m i_m \qquad (L_r \varepsilon k_x) \tag{2}$$

where c is the constant $J(0)$. If J is of degree ν one may lop off in L_r all terms of higher degree than $\nu - \nu_r$ without destroying the equation. If it were possible by some process to change the coefficients L in (2) into invariants, the desired result would follow by induction with respect to the degree of J. In the case of a finite group such a process is readily found: the process of averaging. The linear transformation $U(s)$ of the variables $x_1, \cdots, x_n$ induced by s carries (2) into

$$J = c + L_1^s \cdot i_1 + \cdots + L_m^s \cdot i_m.$$

Summation over s and subsequent division by the number N yields the relation

$$J = c + L_1^* i_1 + \cdots + L_m^* i_m$$

where

$$L_r^* = \frac{1}{N} \cdot \sum_s L_r^s.$$

It is of the same nature as (2), except for the decisive fact that according to their formation the new coefficients L^* are invariants.[2]

Actually Hilbert had to do, not with a finite group, but with the classical problem in which the group Γ consists of all linear transformations s of g variables $\eta_1, \cdots, \eta_g$, and instead of the averaging process he had to resort to a differentiation process invented by Cayley, the so-called Cayley Ω-process, which he skillfully adapted for his end. (It is essential in Cayley's process that the g^2 components of the matrix s are independent variables; instead of the absolutely invariant polynomials J one has to consider

[2] The example of finite groups is used here as an illustration only. Indeed, a direct elementary proof of the first main theorem for finite groups that makes no use of Hilbert's principle (A) has been given by E. Noether, Math. Ann. vol. 77 (1916) p. 89. In dividing by N we have assumed the field k to be of characteristic zero.

relatively invariant homogeneous *forms* each of which has a definite degree and weight.)

Hilbert's theorem (A) is the foundation stone of the general theory of *algebraic manifolds*. Let us now think of k more specifically as the field of all complex numbers. It seems natural to define an algebraic manifold in the space of n coordinates $x_1, \cdots, x_n$ by a number of simultaneous algebraic equations $f_1 = 0, \cdots, f_h = 0$ ($f_i \varepsilon k_x$). According to Theorem (A), nothing would be gained by admitting an infinite number of equations. Let us denote by $Z(f_1, \cdots, f_h)$ the set of points $x = (x_1, \cdots, x_n)$ where $f_1, \cdots, f_h$ and hence all elements of the ideal $\mathcal{J} = \{f_1, \cdots, f_h\}$ vanish simultaneously. g vanishes on $Z(f_1, \cdots, f_h)$ whenever $g \varepsilon \{f_1, \cdots, f_h\}$, but the converse is not generally true. For instance x_1 vanishes wherever x_1^3 does, and yet x_1 is not of the form $x_1^3 \cdot q(x_1 \cdots x_n)$. The language of the algebraic geometers distinguishes here between the simple plane $x_1 = 0$ and the triple plane, although the point set is the same in both cases. Hence what they actually mean by an algebraic manifold is the polynomial ideal and not the point set of its zeros. But even if one cannot expect that every polynomial g vanishing on $Z(f_1, \cdots, f_h) = Z(\mathcal{J})$ is contained in the ideal $\mathcal{J} = \{f_1, \cdots, f_h\}$ one hopes that at least some *power* of g will be. Hilbert's "*Nullstellensatz*" states that this is true, at least if k is the field of complex numbers. It holds in an arbitrary coefficient field k provided one admits points x the coordinates x_i of which are taken from k *or any algebraic extension of* k. Clearly this Nullstellensatz goes to the root of the very concept of algebraic manifolds.[3]

Actually Hilbert conceived it as a tool for the investigation of invariants. As we are now dealing with the full linear group let us consider homogeneous invariants only and drop the adjective homogeneous. Exclude the constants (the invariants of degree 0). Suppose we have ascertained μ non-constant invariants $J_1, \cdots, J_\mu$ such that every non-constant invariant vanishes wherever they vanish simultaneously. An ideal basis of the set $\mathfrak{J}$ of all non-constant invariants certainly meets the demand, but a system $J_1, \cdots, J_\mu$ may be had much more cheaply. Indeed, by a beautiful combination of ideas Hilbert proves that if for a given point $x = x^0$ there exists at all an invariant which neither vanishes for $x = x^0$ nor reduces to a constant, then there exists such an invariant whose weight does not exceed a certain *a priori* limit W (for example, $W = 9n(3n + 1)^8$ for a ternary ground form of

[3] B. L. van der Waerden's book *Moderne Algebra*, vol. 2, 2nd ed., 1940, gives on pp. 1—72 an excellent account of the general algebraic concepts and facts with which we are here concerned.

order n). Hence the $J_1, \cdots, J_\mu$ may be chosen from the invariants of weight not greater than W, and they thus come within the grasp of explicit algebraic construction.

When Hilbert published his proof for the existence of a finite ideal basis, Gordan the formalist, at that time looked upon as the king of invariants, cried out: "This is not mathematics, it is theology!" Hilbert remonstrated then, as he did all his life, against the disparagement of existential arguments as "theology," but we see how, by digging deeper, he was able to meet Gordan's constructive demands. By combining the Nullstellensatz with the Cayley process he further showed that every invariant J is an integral *algebraic* (though not an integral *rational*) function of $J_1, \cdots, J_\mu$, satisfying an equation

$$J^e + G_1 J^{e-1} + \cdots + G_e = 0$$

in which the G's are polynomials of $J_1, \cdots, J_\mu$. Hence it must be possible by suitable algebraic extensions to pass from $J_1, \cdots, J_\mu$ to a full integrity basis. From there on familiar algebraic patterns such as were developed by Kronecker and as are amenable to explicit construction may be followed.

After the formal investigations from Cayley and Sylvester to Gordan, Hilbert inaugurated a new epoch in the theory of invariants. Indeed, by discovering new ideas and introducing new powerful methods he not only brought the subject up to the new level set for algebra by Kronecker and Dedekind, but made such a thorough job of it that he all but finished it, at least as far as the full linear group is concerned. With justifiable pride he concludes his paper, *Ueber die vollen Invariantensysteme*, with the words: "Thus I believe the most important goals of the theory of the function fields generated by invariants have been attained," and therewith quits the scene.[4]

Of later developments which took place after Hilbert quit, two main lines seem to me the most important: (1) The averaging process, which we applied above to finite groups, carries over to continuous compact groups. By this transcendental process of integration over the group manifold, Adolf Hurwitz treated the real orthogonal group. The method has been of great fertility. The simple remark that invariants for the real orthogonal group are *eo ipso* also invariant under the full complex orthogonal group

[4] I recommend to the reader's attention a brief résumé of his invariant-theoretic work which Hilbert himself wrote for the International Mathematical Congress held at Chicago in conjunction with the World Fair in 1893; Collected Papers, vol. 2, item 23.

indicates how the results can be transferred even to non-compact groups, in particular, as it turns out, to all semi-simple Lie groups. (2) Today the theory of invariants for arbitrary groups has taken its natural place within the frame of the theory of representations of groups by linear substitutions, a development which owes its greatest impulse to G. Frobenius.

Although the first main theorem has been proved for wide classes of groups Γ we do not yet know whether it holds for every group. Such attempts as have been made to establish it in this generality were soon discovered to have failed. A promising line for an algebraic attack is outlined in item 14 of Hilbert's Paris list of Mathematical Problems.

Having dwelt in such detail on Hilbert's theory of invariants, we must be brief with regard to his other, more isolated contributions to algebra. The first paper in which the young algebraist showed his real mettle concerns the conditions under which a form with real coefficients is representable as a sum of squares of such forms, in particular with the question whether the obviously necessary condition that the form be positive for all real values of its arguments is sufficient. By ingenious continuity arguments and algebraic constructions Hilbert finds three special cases for which the answer is affirmative, among them of course the positive definite quadratic forms, but counter-examples for all other cases. Similar methods recur in two papers dealing with the attractive problem of the maximum number of real ovals of an algebraic curve or surface and their mutual position. Hilbert conjectured that, irrespective of the number of variables, every *rational function* with real (or rational) coefficients is a sum of squares of such functions provided its values are positive for real values of the arguments; and in his *Grundlagen der Geometrie* he pointed out the role of this fact for the geometric constructions with ruler and "Eichmass." Later O. Veblen conceived, as the basis of the distinction between positive and negative in any field, the axiom that no square sum equals zero. Independently of him, E. Artin and O. Schreier developed a detailed theory of such "real fields," and by means of it Artin succeeded in proving Hilbert's conjecture.[5]

In passing I mention Hilbert's irreducibility theorem according to which one may substitute in an irreducible polynomial suitable integers for all of the variables but one without destroying the irreducibility of the polynomial, and his paper on the solution of the equation of ninth degree by

---

[5] O. Veblen, Trans. Amer. Math. Soc. vol. 7 (1906) pp. 197—199. E. Artin and O. Schreier, Abh. Math. Sem. Hamburgischen Univ. vol. 5 (1926) pp. 85—99; E. Artin, ibid. pp. 100—115.

functions with a minimum number of arguments. They became points of departure for much recent algebraic work (E. Noether, N. Tschebotareff and others). Finally, it ought to be recorded that on the foundations laid by Hilbert a detailed theory of polynomial ideals was erected by E. Lasker and F. S. Macaulay which in turn gave rise to Emmy Noether's general axiomatic theory of ideals. Thus in the field of algebra, as in all other fields, Hilbert's conceptions proved of great consequence for the further development.

Algebraic Number Fields

When Hilbert, after finishing off the invariants, turned to the theory of *algebraic number fields*, the ground had been laid by Dirichlet's analysis of the group of units more than forty years before, and by Kummer's, Dedekind's and Kronecker's introduction of ideal divisors. The theory deals with an algebraic field $\varkappa$ over the field k of rational numbers. One of the most important general results beyond the foundations had been discovered by Dedekind, who showed that the rational prime divisors of the discriminant of $\varkappa$ are at the same time those primes whose ideal prime factors in $\varkappa$ are not all distinct (ramified primes). l being a rational prime, the adjunction to $\varkappa$ of the lth root of a number α in $\varkappa$ yields a relative cyclic field $K = \varkappa(\alpha^{1/l})$ of degree l over $\varkappa$, *provided* $\varkappa$ *contains the* l*th root of unity* $\zeta = e^{2\pi i/l}$ (and according to Lagrange, the most general relative cyclic field of degree l over $\varkappa$ is obtained in this fashion). It may be said that it was this circumstance which forced Kummer, as he tried to prove Fermat's theorem of the impossibility of the equation $\alpha^l + \beta^l = \gamma^l$, to pass from the rational ground field k to the cyclotomic field $\varkappa_l = k(\zeta)$ and then to conceive his ideal numbers in $\varkappa_l$ and to investigate whether the number of classes of equivalent ideal numbers in $\varkappa_l$ is prime to l. Hilbert sharpened his tools in resuming Kummer's study of the relative cyclic fields of degree l over $\varkappa_l$ which he christened "Kummer fields."

His own first important contribution was a theory of *relative Galois fields* K over a given algebraic number field $\varkappa$. His main concern is the relation of the Galois group Γ of $K/\varkappa$ to the way in which the prime ideals of $\varkappa$ decompose in K. Given a prime ideal $\mathfrak{P}$ in K of relative degree f, those substitutions s of Γ for which $s\mathfrak{P} = \mathfrak{P}$ form the splitting group. As always in Galois theory one constructs the corresponding subfield of $K/\varkappa$ (splitting field), to which a number of K belongs if it is invariant under all substitutions of the splitting group. The substititions t which carry every integer A in K

into one, tA, that is congruent to A mod $\mathfrak{P}$ form an invariant subgroup of the splitting group of index f, called the inertial group, and the corresponding field (inertial field) is sandwiched in between the splitting field and K. Let $\mathfrak{p}$ be the prime ideal in $\varkappa$ into which $\mathfrak{P}$ goes, and $\mathfrak{P}^e$ the exact power of $\mathfrak{P}$ by which $\mathfrak{p}$ is divisible. I indicate the nature of Hilbert's results by the following central theorem of his: In the splitting field of $\mathfrak{P}$ the prime ideal $\mathfrak{p}$ in $\varkappa$ splits off the prime factor $\mathfrak{p}^* = \mathfrak{P}^e$ of degree 1 (therefore the name!); in passing from the splitting to the inertial field $\mathfrak{p}^*$ stays prime but its degree increases to f; in passing from the inertial to the full field K, $\mathfrak{p}^*$ breaks up into e equal prime factors $\mathfrak{P}$ of the same degree f. For later application I add the following remarks. If $\mathfrak{P}$ goes into $\mathfrak{p}$ in the first power only, $e = 1$ (which is necessarily so, provided $\mathfrak{p}$ is not a divisor of the relative discriminant of $K/\varkappa$), then the inertial group consists of the identity only. In that case the theory of Galois's strictly finite fields shows that the splitting group is cyclic of order f and that its elements $1, s, s^2, \cdots, s^{f-1}$ are uniquely determined by the congruences

$$sA \equiv A^P, \qquad s^2 A \equiv A^{P^2}, \cdots \qquad (\text{mod } \mathfrak{P})$$

holding for every integer A. Here P is the number of residues in $\varkappa$ modulo $\mathfrak{p}$ and thus P^f the number of residues in K modulo $\mathfrak{P}$. Today we call $s = \sigma(\mathfrak{P})$ the Frobenius substitution of $\mathfrak{P}$; it is of paramount importance that one particular generating substitution of the splitting group may thus be distinguished among all others. One readily sees that for any substitution u of the Galois group $\sigma(u\mathfrak{P}) = u^{-1} \cdot \sigma(\mathfrak{P}) \cdot u$. Thus if the Galois field $K/\varkappa$ is *Abelian*, the substitution $\sigma(\mathfrak{P}) = \sigma(u\mathfrak{P})$ depends on $\mathfrak{p}$ only and may be denoted by $\left(\frac{K}{\mathfrak{p}}\right)$.

In 1893 the Deutsche Mathematiker-Vereinigung asked Hilbert and Minkowski to submit within two years a report on number theory. Minkowski dropped out after a while. Hilbert's monumental report *Die Theorie der algebraischen Zahlkörper* appeared in the Jahresberichte of 1896 (the preface is dated April 1897). What Hilbert accomplished is infinitely more than the Vereinigung could have expected. Indeed, his report is a jewel of mathematical literature. Even today, after almost fifty years, a study of this book is indispensable for anybody who wishes to master the theory of algebraic numbers. Filling the gaps by a number of original investigations, Hilbert welded the theory into an imposing unified body. The proofs of all known theorems he weighed carefully before he decided in favor of those "the principles of which are capable of generalization and the most useful for further research." But before such a selection could be made that "further research" had to be carried out! Meticulous care was given to the notations, with the result that they have been universally adopted (including, to the American printer's dismay, the German letters for ideals). He greatly simplified Kummer's theory, which rested on very complicated calculations, and he introduced those concepts and proved a number of those theorems in which we see today the foundations of a general theory of relative Abelian fields. The most important concept is the norm residue symbol, a pivotal theorem on relative cyclic fields, his famous Satz 90 (Collected Works, vol. I, p. 149). From the preface in which he describes the general character of number theory, and the topics covered by his report in particular, let me quote one paragraph:

"The theory of number fields is an edifice of rare beauty and harmony. The most richly executed part of this building, as it appears to me, is the theory of Abelian fields which Kummer by his work on the higher laws of reciprocity, and Kronecker by his investigations on the complex multiplication of elliptic functions, have opened up to us. The deep glimpses into the theory which the work of these two mathematicians affords reveals at the same time that there still lies an abundance of priceless treasures hidden in this domain, beckoning as a rich reward to the explorer who knows the value of such treasures and with love pursues the art to win them."

Hilbert himself was the miner who during the following two years brought to light much of the hidden ore. The analogy with the corresponding problems in the realm of algebraic functions of one variable where Riemann's powerful instruments of topology and Abelian integrals are available was for him a guiding principle throughout (cf. his remarks in item 12 of his Paris Problems). It is a great pleasure to watch how, step by step, advancing from the special to the general, Hilbert evolves the adequate concepts and methods, and the essential conclusions emerge. I mention his great paper dealing with the relative quadratic fields, and his last and most important *Ueber die Theorie der relativ Abelschen Zahlkörper*. On the basis of the examples he carried through in detail, he conceived as by divination and formulated the basic facts about the so-called class fields. Whereas Hilbert's work on invariants was an end, his work on algebraic numbers was a beginning. Most of the labor of such number theorists of the last decades as Furtwängler, Takagi, Hasse, Artin, Chevalley, has been devoted to proving the results anticipated by Hilbert. By deriving from the

ζ-function the existence of certain auxiliary prime ideals, Hilbert had leaned heavily on transcendental arguments. The subsequent development has gradually eliminated these transcendental methods and shown that though they are the fitting and powerful tool for the investigation of the distribution of prime ideals they are alien to the problem of class fields. In attempting to describe the main issues I shall not ignore the progress and simplification due to this later development.

Hilbert's theory of norm residues is based on the following discoveries of his own: (1) he conceived the basic idea and defined the norm residue symbol for all non-exceptional prime spots; (2) he realized the necessity of introducing infinite prime spots; (3) he formulated the general law of reciprocity in terms of the norm symbol; (4) he saw that by means of that law one can extend the definition of the norm symbol to the exceptional prime spots where the really interesting things happen. — It was an essential progress when E. Artin later (5) replaced the roots of unity by the elements of the Galois group as values of the residue symbol. In sketching Hilbert's problems I shall make use of this idea of Artin's and also of the abbreviating language of (6) Hensel's p-adic numbers and (7) Chevalley's *idèles*.[6]

As everybody knows, an integer a indivisible by the prime $p \neq 2$ is said to be a quadratic residue if the congruence $x^2 \equiv a \pmod{p}$ is solvable. Gauss introduced the symbol $\left(\frac{a}{p}\right)$ which has the value $+1$ or -1 according to whether a is a quadratic residue or non-residue mod p, and observed that it is a character, $\left(\frac{a}{p}\right) \cdot \left(\frac{a'}{p}\right) = \left(\frac{aa'}{p}\right)$. Indeed, the p residues modulo p — as whose representatives one may take $0, 1, \cdots, (p-1)$ — form a *field*, and after exclusion of 0 a *group* in which the quadratic residues form a subgroup of index 2. Let $K = k(b^{1/2})$ be a quadratic field which arises from the rational ground field k by adjunction of the square root of the rational number b. An integer $\alpha \neq 0$ is called by Hilbert *a p-adic norm* in K if modulo any given power of p it is congruent to the norm of a suitable integer in K. He sets $\left(\frac{a, K}{p}\right) = +1$ if a is p-adic norm, -1 jn the opposite case, and finds that this p-adic norm symbol again is a character. The investigation of numbers modulo arbitrarily high powers of p was systematized by K. Hensel in the form of his p-adic numbers, and I repeat Hilbert's definition in this language:

---

[6] The latest account of the theory is C. Chevalley's paper *La théorie du corps de classes*, Ann. of Math. vol. 41 (1940) pp. 394—418.

"The rational number $a \neq 0$, or more generally the p-adic number $a_p \neq 0$, is a p-adic norm in K if the equation

$$a_p = \mathrm{Nm}\,(x + yb^{1/2}) = x^2 - by^2$$

has a p-adic solution $x = x_p$, $y = y_p$; the norm symbol (a_p, K) equals $+1$ or -1 according to whether or not a_p is (p-adic) norm in K." The p-adic numbers form a field $k(p)$ and after exclusion of 0 a multiplicative group G_p in which, according to Hilbert's result, the p-adic norms in K form a subgroup of index 2 or 1. The cyclic nature of the factor group is the salient point. One easily finds that the p-adic squares form a subgroup G_p^2 of index 4 if $p \neq 2$, of index 8 if $p = 2$, and thus the factor group G_p/G_p^2 is not cyclic and could not be described by a single character. Of course every p-adic square is a p-adic norm in K. Both steps, the substitution of K-norms for squares and the passage from the modulus p to arbitrarily high powers of p; the first step amounting to a relaxation, the second to a sharpening of Gauss's condition for quadratic residues; are equally significant for the success of Hilbert's definition.

Every p-adic number $a_p \neq 0$ is of the form $p^h \cdot e_p$ where e_p is a p-adic unit, and thus a_p is of a definite *order* h (at p). An ordinary rational number a coincides with a definite p-adic number $I_p(a) = a_p$. Here I_p symbolizes a homomorphic projection of k into $k(p)$:

$$I_p(a + a') = I_p(a) + I_p(a'), \qquad I_p(aa') = I_p(a) \cdot I_p(a').$$

The character $\left(\frac{a, K}{p}\right)$ is identical with $(I_p(a), K)$.

We come to Hilbert's second discovery: he realized that simple laws will not result unless one adds to the "finite prime spots" p one infinite prime spot q. By definition the q-adic numbers are the real numbers and $I_q(a)$ is the real number with which the rational number a coincides. Hence the real number a_q is a q-adic norm in K if the equation $a_q = x^2 - by^2$ has a solution in real numbers x, y. Clearly if $b > 0$ or K is real, this is the case for every a_q; if however $b < 0$ or K is imaginary, only the positive numbers a_q are q-adic norms. Hence

$$(a_q, K) = 1 \text{ if } K \text{ real}; \qquad (a_q, K) = \operatorname{sgn} a_q \text{ if } K \text{ imaginary}.$$

The fact that the norm symbol is a character is thus much more easily verified for the infinite prime spot than for the finite ones.

Hilbert's third observation is to the effect that Gauss's reciprocity law with its two supplements may be condensed into the elegant formula

$$\prod_p (I_p(a), K) = \prod_p \left(\frac{a, K}{p}\right) = 1, \tag{3}$$

the product extending over the infinite and all finite prime spots p. There is no difficulty in forming this product because almost all factors (that is, all factors with but a finite number of exceptions) equal unity. Indeed, if the prime p does not go into the discriminant of K, then $(a_p, K) = 1$ for every p-adic unit a_p. Formula (3) is the first real vindication for the norm residue idea, which must have given Hilbert the assurance that the higher reciprocity laws had to be formulated in terms of norm residues.

A given rational number a assigns to every prime spot p a p-adic number $a_p = I_p(a)$. On which features of this assignment does one rely in forming the product (3)? The obvious answer is given by Chevalley's notion of *idèle*: an idèle $\boldsymbol{a}$ is a function assigning to every prime spot p a p-adic number $a_p \neq 0$ which is a p-adic unit for almost all prime spots p. The idèles form a multiplicative group J_k. By virtue of the assignment $a_p = I_p(a)$ every rational number $a \neq 0$ gives rise to an idèle, called the *principal idèle* a. With the idèles $\boldsymbol{a}$ at our disposal we might as well return to the notation $\left(\frac{\boldsymbol{a}, K}{p}\right)$ for (a_p, K). The formula

$$\phi_K(\boldsymbol{a}) = (\boldsymbol{a}, K) = \prod_p (a_p, K) = \prod_p \left(\frac{\boldsymbol{a}, K}{p}\right) \tag{4}$$

defines a character ϕ_K, the norm character, in the group J_k of all idèles. The reciprocity law in Hilbert's form (3) maintains that

$$(\boldsymbol{a}, K) = 1 \tag{5}$$

if $\boldsymbol{a}$ is principal. By the very definition of the norm symbol (a_p, K) the same equation holds if $\boldsymbol{a}$ is a norm in K, that is, if a_p is a p-adic norm in K for every prime spot p. Two idèles $\boldsymbol{a}, \boldsymbol{a}'$ are said to be equivalent, $\boldsymbol{a} \sim \boldsymbol{a}'$, if their quotient $\boldsymbol{a}'\boldsymbol{a}^{-1}$ is principal. Let us denote by $\mathrm{Nm}J_K$ the group of all idèles which are equivalent to norms in K. Then (5) holds for all idèles $\boldsymbol{a}$ of $\mathrm{Nm}J_K$; it would be good to know that it holds for no other idèles, or, in other words, that $\mathrm{Nm}J_K$ is a subgroup of J_k of index 2.

The stage is now reached where the experiences gathered for a quadratic field K over the rational ground field k may be generalized to any relative Abelian field K over a given algebraic number field $\varkappa = k(\theta)$. First a word about the infinite prime spots of $\varkappa$. The defining equation $f(\theta) = 0$, an irreducible equation in k of some degree m, has m distinct roots $\theta', \theta'', \cdots, \theta^{(m)}$ in the continuum of complex numbers. Suppose that r of them are real, say $\theta', \cdots, \theta^{(r)}$. Then each element α of $\varkappa$ has its r real conjugates $\alpha', \cdots, \alpha^{(r)}$, and $\alpha^{(t)}$ arises from α by a homomorphic projection $I^{(t)}$ of $\varkappa$ into the field of all real numbers,

$$\alpha \to \alpha^{(t)} = I^{(t)}(\alpha) \qquad (t = 1, \cdots, r).$$

We therefore speak of r real infinite prime spots $\mathfrak{q}', \cdots, \mathfrak{q}^{(r)}$ with the corresponding projections $I' = I_{\mathfrak{q}'}, \cdots, I^{(r)}$; the fields $\varkappa(\mathfrak{q}'), \cdots, \varkappa(\mathfrak{q}^{(r)})$ are identical with the field of all real numbers. Thus a is an nth $\mathfrak{q}'$-adic power if the equation $\alpha' = \xi'^n$ has a real solution ξ'. One sees that this imposes a condition only if n is even, and then requires α' to be positive. (In the complex domain the equation is always solvable whether n be even or odd, and that is the reason why we ignore the complex infinite prime spots altogether.)

The finite prime spots are the prime ideals $\mathfrak{p}$ of $\varkappa$. In studying a Galois field $K/\varkappa$ of relative degree n we first exclude the ramified ideals $\mathfrak{p}$ which go into the relative discriminant of $K/\varkappa$. An unramified ideal $\mathfrak{p}$ of $\varkappa$ factors in K into a number g of distinct prime ideals $\mathfrak{P}_1 \cdots \mathfrak{P}_g$ of relative degree f, $fg = n$. It is easily seen that a $\mathfrak{p}$-adic number $\alpha_{\mathfrak{p}} \neq 0$ is a $\mathfrak{p}$-adic norm in K if and only if its order (at $\mathfrak{p}$) is a multiple of f. In particular, the $\mathfrak{p}$-adic units are norms. Thus we encounter a situation which is essentially simpler than the one taken care of by Gauss's quadratic residue symbol: the norm character of $\alpha_{\mathfrak{p}}$ depends only on the order i at $\mathfrak{p}$ of $\alpha_{\mathfrak{p}}$. It is now clear how to proceed: we choose a primitive fth root of unity ζ and define $(\alpha_{\mathfrak{p}}, K) = \zeta^i$ if $\alpha_{\mathfrak{p}}$ is of order i. This function of $\alpha_{\mathfrak{p}} \neq 0$ is a character which assumes the value 1 for the norms and the norms only. But here is the rub: there is no algebraic property distinguishing the several primitive fth roots of unity from one another. Thus the choice of ζ among them remains arbitrary. One could put up with this if one dealt with one prime ideal only. But when one has to take all prime spots simultaneously into account, as is necessary in forming products of the type (4), then the arbitrariness involved in the choice of ζ for each $\mathfrak{p}$ will destroy all hope of obtaining a simple reciprocity law like (5). I shall forego describing the devices by which Eisenstein, Kummer, Hilbert extricated themselves from this entanglement. By far the best solution was found by Artin: if $K/\varkappa$ is Abelian, then the Frobenius substitution $\left(\frac{K}{\mathfrak{p}}\right)$ is uniquely determined by K and $\mathfrak{p}$ and is an element of order f of the Galois group Γ of $K/\varkappa$. Let this element of the Galois group replace ζ in our final definition of the $\mathfrak{p}$-adic norm symbol:

$$(\alpha_{\mathfrak{p}}, K) = \left(\frac{\alpha, K}{\mathfrak{p}}\right) = \left(\frac{K}{\mathfrak{p}}\right)^i \text{ if } \alpha_{\mathfrak{p}} \text{ is of order } i \text{ at } \mathfrak{p}. \tag{6}$$

We could now form for any idèle $\boldsymbol{\alpha}$ the product

$$\prod_{\mathfrak{p}} (\alpha_{\mathfrak{p}}, K) = \prod_{\mathfrak{p}} \left(\frac{\boldsymbol{\alpha}, K}{\mathfrak{p}}\right) = (\boldsymbol{\alpha}, K)$$

extending over all finite and (real) infinite prime spots $\mathfrak{p}$ and formulate the reciprocity law asserting that $(\boldsymbol{\alpha}, K) = 1$ for any principal idèle $\boldsymbol{\alpha}$ – had we not excluded certain exceptional prime spots in our definition of $(\alpha_{\mathfrak{p}}, K)$, namely the infinite prime spots and the ramified prime ideals. In the special case he investigated Kummer had succeeded in obtaining the correct value of $(\alpha_{\mathfrak{p}}, K)$ for the exceptional $\mathfrak{p}$ by extremely complicated calculations. Hilbert's fourth discovery is a simple and ingenious method of circumventing this formidable obstacle which blocked the road to further progress. Let us first restrict ourselves to idèles $\boldsymbol{\alpha}$ which are nth powers at our exceptional prime spots; in other words, we assume that the equation $\alpha_{\mathfrak{p}} = \xi_{\mathfrak{p}}^n$ is solvable for the $\mathfrak{p}$-adic values $\alpha_{\mathfrak{p}}$ of $\boldsymbol{\alpha}$ at this finite number of prime spots. There is no difficulty in defining $(\boldsymbol{\alpha}, K)$ under this restriction:

$$(\boldsymbol{\alpha}, K) = \prod_{\mathfrak{p}}{}' (\alpha_{\mathfrak{p}}, K),$$

the product extending, as indicated by the accent, over the non-exceptional prime spots only, for which we know what $(\alpha_{\mathfrak{p}}, K)$ means. Under the same restriction we prove (with Artin) the reciprocity law

$$(\boldsymbol{\alpha}, K) = 1 \text{ if } \boldsymbol{\alpha} \text{ is principal}, \tag{7}$$

and observe that by its very definition $(\boldsymbol{\alpha}, K) = 1$ if $\boldsymbol{\alpha}$ is norm. We now return to an *arbitrary* idèle $\boldsymbol{\alpha}$. It is easily shown that there exists an equivalent idèle $\boldsymbol{\alpha}^* \sim \boldsymbol{\alpha}$ which is an nth power at all exceptional prime spots, but of course there will be plenty of them. However, the restricted law of reciprocity insures that

$$(\boldsymbol{\alpha}^*, K) = \prod_{\mathfrak{p}}{}'(\alpha_{\mathfrak{p}}^*, K)$$

has the same value for every one of the $\boldsymbol{\alpha}^*$'s, and it is this value which we now denote by $(\boldsymbol{\alpha}, K)$. This definition adopted, the reciprocity law (7) and the statement that $(\boldsymbol{\alpha}, K) = 1$ for every norm α follow at once *without restriction.* Thus the reciprocity law itself is made the tool for getting the exceptional prime spots under control!

Once $(\boldsymbol{\alpha}, K)$ is known for every idèle $\boldsymbol{\alpha}$ we can compute $(\alpha_{\mathfrak{p}}, K)$ for a given prime spot $\mathfrak{p}$ and a given $\mathfrak{p}$-adic number $\alpha_{\mathfrak{p}} \neq 0$ as the value of $(\boldsymbol{\alpha}, K)$ for that "primary" idèle, also denoted by $\alpha_{\mathfrak{p}}$, which equals $\alpha_{\mathfrak{p}}$ at $\mathfrak{p}$ and 1 at any other prime spot. (The idèle $\boldsymbol{\alpha}$ is the product of its primary components, $\boldsymbol{\alpha} = \Pi_{\mathfrak{p}}\alpha_{\mathfrak{p}}$.) One expects the following two propositions to hold:

I. $(\alpha_{\mathfrak{p}}, K) = 1$ *if and only if* $\alpha_{\mathfrak{p}}$ *is a* $\mathfrak{p}$*-adic norm.*

II. *Given a prime ideal* $\mathfrak{p}$, $(\alpha_{\mathfrak{p}}, K) = 1$ *for every* $\mathfrak{p}$*-adic unit* $\alpha_{\mathfrak{p}}$ *if and only if* $\mathfrak{p}$ *is unramified.*

The direct parts of I and II:

(I_0) $\alpha_{\mathfrak{p}} =$ norm inplies $(\alpha_{\mathfrak{p}}, K) = 1$;

(II_0) $\mathfrak{p}$ unramified implies $(\alpha_{\mathfrak{p}}, K) = 1$ for every $\mathfrak{p}$-adic unit $\alpha_{\mathfrak{p}}$, were settled above. The converse statement of I_0 is trivial for the non-exceptional prime spots; but owing to the indirect definition of the norm symbol for the exceptional prime spots, the proofs of the converse of I_0 for the exceptional spots and of the converse of II_0 are rather intricate. From II we learn that for none of the ramified prime ideals $\mathfrak{p}$ does the norm character of $\alpha_{\mathfrak{p}}$ depend on the order of $\alpha_{\mathfrak{p}}$ only: this simple feature which makes the definition (6) possible is limited to non-ramified $\mathfrak{p}$. One would also expect:

III. *If the principal idèle* $\boldsymbol{\alpha}$ *is an idèle norm in* K, *then the number* α *is norm of a number in* K.

This is true for *cyclic* fields $K/\varkappa$, but in general not for Abelian fields.

Let us again denote by $\mathrm{Nm}J_K$ the subgroup in $J_\varkappa$ of the idèles which are equivalent to norms. Then the norm symbol $\phi_K(\boldsymbol{\alpha}) = (\boldsymbol{\alpha}, K)$ determines a homomorphic mapping of the factor group $J_\varkappa/\mathrm{Nm}J_K$ into the Galois group of $K/\varkappa$. One would expect that this mapping is one-to-one:

IV. *By means of the norm symbol the factor group* $J_\varkappa/\mathrm{Nm}J_K$ *is isomorphically mapped onto the Galois group of* $K/\varkappa$.

I, II, III_c (the subscript c indicating restriction to cyclical fields) and IV are the main propositions of what one might call the *norm theory of relative Abelian fields*. They refer to a *given* field $K/\varkappa$.

There is a second part of the theory, *the class field theory proper*, which is concerned with the manner in which *all possible* relative Abelian fields K over $\varkappa$ are reflected in the structure of the group $J_\varkappa$ of idèles in $\varkappa$. Each such field K determines, as we have seen, a subgroup $\mathrm{Nm}J_K$ of $J_\varkappa$ of finite index. The question arises *which* subgroups $J_\varkappa^*$ of $J_\varkappa$ are generated in this way by Abelian fields $K/\varkappa$. Clearly the following conditions are necessary:

(1) Every principal idèle is in $J_\varkappa^*$.

(2) There is a natural number n such that every nth power of an idèle is in $J_\varkappa^*$.

(3) There is a finite set S of prime spots such that $\boldsymbol{\alpha}$ is in $J_\varkappa^*$ provided $\boldsymbol{\alpha}$ is a unit at every prime spot and equals 1 at the prime spots of S.

The main theorem concerning class fields states that these conditions are also sufficient.

V. *Given a subgroup* $J_\varkappa^*$ *of* $J_\varkappa$ *fulfilling the above three conditions* (and therefore, as one readily verifies, of finite index), *there exists a uniquely determined Abelian field* $K/\varkappa$ *such that* $J_\varkappa^* = \mathrm{Nm}J_K$.

We divide the idèles of $\varkappa$ into *classes* by throwing two idèles into the same class if their quotient is in $J_\varkappa^*$. Then $J_\varkappa/J_\varkappa^*$ is the class group and K is called the corresponding *class field*. The most important example results if one lets $J_\varkappa^*$ consist of the *unit idèles* $\boldsymbol{\alpha}$ whose values $\alpha_\mathfrak{p}$ are $\mathfrak{p}$-adic units at every prime spot $\mathfrak{p}$.[7] Then the classes may be described as the familiar classes of *ideals*: two ideals are put in the same class if their quotient is a principal ideal (α) springing from a number α positive at all real infinite prime spots. The corresponding class field K, the so-called absolute class field, is of relative discriminant 1, and the largest unramified Abelian field over $\varkappa$ (Theorem II). Its degree n over $\varkappa$ is the class number of ideals, its Galois group isomorphic to the class group of ideals in $\varkappa$ (Theorem IV). f being the least power of $\mathfrak{p}$ which lies in the principal class, $\mathfrak{p}$ decomposes into n/f distinct prime ideals in K, each of relative degree f. This last statement does nothing but repeat the norm definition of the class field. Hence the way in which $\mathfrak{p}$ factors in K depends only on the class to which $\mathfrak{p}$ belongs. The easiest way of extending the theory from the case with no ramified prime ideals, which was preponderant in Hilbert's thought, to Takagi's ramified

---

[7] At the (real) infinite prime spots the positive numbers are considered the units.

case is by substituting idèles for ideals. Hilbert also stated that every ideal in $\varkappa$ becomes a principal ideal in the absolute class field. It is today possible to show that this is so, by arguments, however, which are far from being fully understood, because this question transcends the domain of Abelian fields.

As was stated above, Hilbert did not prove these theorems in their full generality, but taking his departure from Gauss's theory of genera in quadratic fields and Kummer's investigations he worked his way gradually up from the simplest cases, developing as he went along the necessary machinery of new concepts and propositions about them until he could survey the whole landscape of class fields. We cannot attempt here to give an idea of the highly involved proofs. The completion of the work he left to his successors. The day is probably still far off when we shall have a theory of relative *Galois* number fields of comparable completeness.

Kronecker had shown, and Hilbert found a simpler proof for the fact, that Abelian fields over the rational ground field k are necessarily subfields of the cyclotomic fields, and are thus obtained from the transcendental function $e^{2\pi i x}$ by substituting rational values for the argument x. For Abelian fields over an imaginary quadratic field the so-called complex multiplication of the elliptic and modular functions plays a similar role ("Kronecker's Jugendtraum"). While Heinrich Weber following in Kronecker's footsteps, and R. Fuëter under Hilbert's guidance, made this dream come true, Hilbert himself began to play with modular functions of several variables which are defined by means of algebraic number fields, and to study their arithmetical implications. He never published these investigations, but O. Blumenthal, and later E. Hecke, used his notes and developed his ideas. The results are provocative, but still far from complete. It is indicative of the fertility of Hilbert's mind at this most productive period of his life that he handed over to his pupils a complex of problems of such fascination as that of the relation between number theory and modular functions.[8]

There remain to be mentioned a particularly simple proof of the transcendence of e and π with which Hilbert opened the series of his arithmetical papers, and the 1909 paper settling Waring's century-old conjecture. I

---

[8] R. Fuëter, *Singuläre Moduln und complexe Multiplication*, 2 vols., Leipzig, 1924, 1927; cf. also H. Hasse, J. Reine Angew. Math. vol. 157 (1927) pp. 115—139. O. Blumenthal, Math. Ann. vol. 56 (1903) pp. 509—548, vol. 58 (1904) pp. 497—527. E. Hecke, Math. Ann. vol. 71 (1912) pp. 1—37, vol. 74 (1913) pp. 465—510.

should classify the latter paper among his most original ones, but we can forego considering it more closely because a decade later Hardy and Littlewood found a different approach which yields asymptotic formulas for the number of representations, and it is the Hardy-Littlewood "circle method" which has given rise in recent times to a considerable literature on this and related subjects.[9]

Axiomatics

There could not have been a more complete break than the one dividing Hilbert's last paper on the theory of number fields from his classical book, *Grundlagen der Geometrie*, published in 1899. Its only forerunner is a note of the year 1895 on the straight line as the shortest way. But O. Blumenthal records that as early as 1891 Hilbert, discussing a paper on the role of Desargues's and Pappus's theorems read by H. Wiener at a mathematical meeting, made a remark which contains the axiomatic standpoint in a nutshell: "It must be possible to replace in all geometric statements the words *point*, *line*, *plane* by *table*, *chair*, *mug*."

The Greeks had conceived of geometry as a deductive science which proceeds by purely logical processes once the few axioms have been established. Both Euclid and Hilbert carry out this program. However, Euclid's list of axioms was still far from being complete; Hilbert's list is complete and there are no gaps in the deductions. Euclid tried to give a descriptive definition of the basic spatial objects and relations with which the axioms deal; Hilbert abstains from such an attempt. All that we must know about those basic concepts is contained in the axioms. The axioms are, as it were, their implicit (necessarily incomplete) definitions. Euclid believed the axioms to be evident; his concern is the real space of the physical universe. But in the deductive system of geometry the evidence, even the truth of the axioms, is irrelevant; they figure rather as hypotheses of which one sets out to develop the logical consequences. Indeed there are many different material interpretations of the basic concepts for which the axioms become true. For instance, the axioms of n-dimensional Euclidean vector geometry hold if a distribution of direct current in a given electric circuit, the n branches of which connect in certain branching points, is called a vector, and Joule's heat produced per unit time by the current is considered the

[9] It must suffice here to quote the first paper in this line: G. H. Hardy and J. E. Littlewood, Quart. J. Math. vol. 48 (1919) pp. 272—293, and its latest successor which carries Waring's theorem over to arbitrary algebraic fields: C. L. Siegel, Amer. J. Math. vol. 66 (1944) pp. 122—136.

square of the vector's length. In building up geometry on the foundation of its axioms one will attempt to economize as much as possible and thus illuminate the role of the several groups of axioms. Arranged in their natural hierarchy they are the axioms of incidence, order, congruence, parallelism, and continuity. For instance, if the theory of geometric proportions or of the areas of polygons can be established without resorting to the axioms of continuity, this ought to be done.

In all this, though the execution shows the hand of a master, Hilbert is not unique. An outstanding figure among his predecessors is M. Pasch, who had indeed travelled a long way from Euclid when he brought to light the hidden axioms of order and with methodical clarity carried out the deductive program for projective geometry (1882). Others in Germany (F. Schur) and a flourishing school of Italian geometers (Peano, Veronese) had taken up the discussion. With respect to the economy of concepts, Hilbert is more conservative than the Italians: quite deliberately he clings to the Euclidean tradition with its three classes of undefined elements, points, lines, planes, and its relations of incidence, order and congruence of segments and angles. This gives his book a peculiar charm, as if one looked into a face thoroughly familiar and yet sublimely transfigured.

It is one thing to build up geometry on sure foundations, another to inquire into the logical structure of the edifice thus erected. If I am not mistaken, Hilbert is the first who moves freely on this higher "metageometric" level: systematically he studies the mutual independence of his axioms and settles the question of independence from certain limited groups of axioms for some of the most fundamental geometric theorems. His method is the *construction of models*: the model is shown to disagree with one and to satisfy all other axioms; hence the one cannot be a consequence of the others. One outstanding example of this method had been known for a considerable time, the Cayley-Klein model of non-Euclidean geometry. For Veronese's non-Archimedean geometry Levi-Civita (shortly before Hilbert) had constructed a satisfactory arithmetical model. The question of *consistency* is closely related to that of independence. The general ideas appear to us almost banal today, so thoroughgoing has been their influence upon our mathematical thinking. Hilbert stated them in clear and unmistakable language, and embodied them in a work that is like a crystal: an unbreakable whole with many facets. Its artistic qualities have undoubtedly contributed to its success as a masterpiece of science.

In the construction of his models Hilbert displays an amazing wealth of invention. The most interesting examples seem to me the one by which he shows that Desargues's theorem does not follow from the plane incidence axioms, but that the plane incidence axioms combined with Desargues's theorem enable one to embed the plane in a higher dimensional space in which all incidence axioms hold; and then the other example by which he decides whether the Archimedean axiom of continuity is necessary to restore the full congruence axioms after having curtailed them by the exclusion of reflections.

What is the building material for the models? Klein's model of non-Euclidean geometry could be interpreted as showing that he who accepts Euclidean geometry with its points and lines, and so on, can by mere change of nomenclature also get the non-Euclidean geometry. Klein himself preferred another interpretation in terms of projective space. However, Descartes's analytic geometry had long provided a more general and satisfactory answer, of which Riemann, Klein and many others must have been aware: All that we need for our construction is the field of real *numbers*. Hence any contradiction in Euclidean geometry must show up as a contradiction in the arithmetical axioms on which our operations with real numbers are based. Nobody had said that quite clearly before Hilbert. He formulates a complete and simple set of axioms for real numbers. The system of arithmetical axioms will have its exchangeable parts just as the geometric system has. From a purely algebraic standpoint the most important axioms are those characterizing a (commutative or non-commutative) *field*. Any such abstract number field may serve as a basis for the construction of corresponding geometries. *Vice versa*, one may introduce numbers and their operations in terms of a space satisfying certain axioms; Hilbert's Desarguesian *Streckenrechnung* is a fine example. In general this reverse process is the more difficult one. The Chicago school under E. H. Moore took up Hilbert's investigations, and in particular O. Veblen did much to reveal the perfect correspondence between the projective spaces obeying a set of simple incidence axioms (and no axioms of order), and the abstractly defined number fields.[10]

What the question of independence literally asks is to make sure that a certain proposition cannot be deduced from other propositions. It seems to require that we make the propositions, rather than the things of which

---

[10] Among later contributions to this question I mention W. Schwan, *Streckenrechnung und Gruppentheorie*, Math. Zeit. vol. 3 (1919) pp. 11—28. A complete bibliography of geometric axiomatics since Hilbert would probably cover many pages. I refrain from citing a list of names.

they speak, the object of our investigation, and that as a preliminary we fully analyze the logical mechanism of deduction. The method of models is a wonderful trick to avoid that sort of logical investigations. It pays, however, a heavy price for thus shirking the fundamental issue: it merely reduces everything to the question of consistency for the arithmetical axioms, which is left unanswered. In the same manner *completeness*, which literally means that every general proposition about the objects with which the axioms deal can be decided by inference from the axioms, is replaced by *categoricity* (Veblen), which asserts that any imaginable model is isomorphic to the one model by which consistency is established. In this sense Hilbert proves that there is but "one," the Cartesian geometry, which fulfills all his axioms. Only in the case of G. Fano's and O. Veblen's finite projective spaces, for example, of the projective plane consisting of seven points, the model is a purely combinatorial scheme, and the questions of consistency, independence and completeness can be answered in the absolute sense. Hilbert never seems to have thought of illustrating his conception of the axiomatic method by purely combinatorial schemes, and yet they provide by far the simplest examples.

An approach to the foundations of geometry entirely different from the one followed in his book is pursued by Hilbert in a paper which is one of the earliest documents of set-theoretic topology. From the standpoint of mechanics the central task which geometry ought to perform is that of describing the mobility of a solid. This was the viewpoint of Helmholtz, who succeeded in characterizing the group of motions in Euclidean space by a few simple axioms. The question had been taken up by Sophus Lie in the light of his general theory of continuous groups. Lie's theory depends on certain assumptions of differentiability; to get rid of them is one of Hilbert's Paris Problems. In the paper just mentioned he *does* get rid of them as far as Helmholtz's problem in the plane is concerned. The proof is difficult and laborious; naturally continuity is now the foundation — and not the key-stone of the building as it had been in his *Grundlagen* book. Other authors, R. L. Moore, N. J. Lennes, W. Süss, B. v. Kérékjarto, carried the problem further along these topological lines. A half-personal reminiscence may be of interest. Hilbert defines a two-dimensional manifold by means of neighborhoods, and requires that a class of "admissible" one-to-one mappings of a neighborhood upon Jordan domains in an x, y-plane be designated, any two of which are connected by continuous transformations. When I gave a course on Riemann surfaces at Göttingen in 1912, I consulted Hilbert's paper and noticed that the neighborhoods themselves could be used to characterize that class. The ensuing definition was given its final touch by F. Hausdorff; the Hausdorff axioms have become a byword in topology.[11] (However, when it comes to explaining what a *differentiable* manifold is, we are to this day bound to Hilbert's roundabout way; cf. Veblen and Whitehead, *The foundations of differential geometry*, Cambridge, 1932.)

The fundamental issue of an *absolute proof of consistency* for the axioms which should include the whole of mathematical analysis, nay even Cantor's set theory in its wildest generality, remained in Hilbert's mind, as a paper read before the International Congress at Heidelberg in 1904 testifies. It shows him on the way, but still far from the goal. Then came the time when integral equations and later physics became his all-absorbing interest. One hears a loud rumbling of the old problem in his Zürich address, *Axiomatisches Denken*, of 1917. Meanwhile the difficulties concerning the foundations of mathematics had reached a critical stage, and the situation cried for repair. Under the impact of undeniable antinomies in set theory, Dedekind and Frege had revoked their own work on the nature of numbers and arithmetical propositions, Bertrand Russell had pointed out the hierarchy of types which, unless one decides to "reduce" them by sheer force, undermine the arithmetical theory of the continuum; and finally L. E. J. Brouwer by his intuitionism had opened our eyes and made us see how far generally accepted mathematics goes beyond such statements as can claim real meaning and truth founded on evidence. I regret that in his opposition to Brouwer, Hilbert never openly acknowledged the profound debt which he, as well as all other mathematicians, owes Brouwer for this revelation.

Hilbert was not willing to make the heavy sacrifices which Brouwer's standpoint demanded, and he saw, at least in outline, a way by which the cruel mutilation could be avoided. At the same time he was alarmed by signs of wavering loyalty within the ranks of mathematicians, some of whom openly sided with Brouwer. My own article on the *Grundlagenkrise* in Math. Zeit. vol. 10 (1921), written in the excitement of the first postwar years in Europe, is indicative of the mood. Thus Hilbert returns to the problem of foundations in earnest. He is convinced that complete certainty can be restored without "committing treason to our science." There is anger and determination in his voice when he proposes "die Grundlagenfragen einfürallemal aus der Welt zu schaffen." "Forbidding a mathematician to make

---

[11] A parallel development, with E. H. Moore as the chief prompter, must have taken place in this country. As I have to write from memory mainly, it is inevitable that my account should be colored by the local Göttingen tradition.

use of the principle of excluded middle," says he, "is like forbidding an astronomer his telescope or a boxer the use of his fists."

Hilbert realized that the mathematical statements themselves could not be made the subject of a mathematical investigation whose aim is to answer the question of their consistency in its primitive sense, unless they be first reduced to mere *formulas*. Algebraic formulas like $a + b = b + a$ are the most familiar examples. The process of deduction by which formulas previously obtained give rise to new formulas must be described without reference to any meaning of the formulas. The deduction starts from certain initial formulas, the axioms, which must be written out explicitly. Whereas in his *Grundlagen der Geometrie* the meaning of the geometric terms had become irrelevant, but the meaning of logical terms, as "and," "not," "if then," had still to be understood, now every trace of meaning is obliterated. As a consequence, logical symbols like $\rightarrow$ in $a \rightarrow b$, read: a implies b, enter into the formulas. Hilbert fully agrees with Brouwer in that the great majority of mathematical propositions are not "real" ones conveying a definite meaning verifiable in the light of evidence. But he insists that the non-real, the "ideal propositions" are indispensable in order to give our mathematical system "completeness." Thus he parries Brouwer, who had asked us to give up what is meaningless, by relinquishing the pretension of meaning altogether, and what he tries to establish is not *truth* of the individual mathematical proposition, but *consistency* of the system. The game of deduction when played according to rules, he maintains, will never lead to the formula $0 \neq 0$. In this sense, and in this sense only, he promises to salvage our cherished classical mathematics in its entirety.

For those who accuse him of degrading mathematics to a mere game he points first to the introduction of ideal elements for the sake of completeness as a common method in all mathematics -- for example, of the ideal points outside an accessible portion of space, without which space would be incomplete —; secondly, to the neighboring science of physics where likewise, not the individual statement is verifiable by experiment, but in principle only the system as a whole can be confronted with experience.

But how to make sure that the "game of deduction" never leads to a contradiction? Shall we prove this by the same mathematical method the validity of which stands in question, namely by deduction from axioms? This would clearly involve a regress *ad infinitum*. It must have been hard on Hilbert, the axiomatist, to acknowledge that the insight of consistency is rather to be attained by *intuitive reasoning* which is based on evidence and not on axioms. But after all, it is not surprising that ultimately the mind's seeing eye must come in. Already in communicating the rules of the game we must count on understanding. The game is played in silence, but the rules must be *told* and any reasoning about it, in particular about its consistency, communicated by *words*. Incidentally, in describing the indispensable intuitive basis for his *Beweistheorie* Hilbert shows himself an accomplished master of that, alas, so ambiguous medium of communication, language. With regard to what he accepts as evident in this "metamathematical" reasoning, Hilbert is more papal than the pope, more exacting than either Kronecker or Brouwer. But it cannot be helped that our reasoning in following a hypothetic sequence of formulas leading up to the formula $0 \neq 0$ is carried on in hypothetic generality and uses that type of evidence which a formalist would be tempted to brand as application of the principle of complete induction. Elementary arithmetics can be founded on such intuitive reasoning as Hilbert himself describes, but we need the formal apparatus of variables and "quantifiers" to invest the infinite with the all important part that it plays in higher mathematics. Hence Hilbert prefers to make a clear cut: he becomes strict formalist in mathematics, strict intuitionist in metamathematics.

It is perhaps possible to indicate briefly how Hilbert's formalism restores the *principle of the excluded middle* which was the main target of Brouwer's criticism. Consider the infinite sequence of numbers 0, 1, 2, $\cdots$. Any property A of numbers (for example, "being prime") may be represented by a propositional function $A(x)$ ("x is prime"), from which a definite proposition $A(b)$ arises by substituting a concrete number b for the variable x ("6 is prime"). Accepting the principle which Brouwer denies and to which Hilbert wishes to hold on, that (i) either there exists a number x for which $A(x)$ holds, or (ii) $A(x)$ holds for no x, we can find a "representative" r for the property A, a number such that $A(b)$ implies $A(r)$ whatever the number b, $A(b) \rightarrow A(r)$. Indeed, in the alternative (i) we choose r as one of the numbers x for which $A(x)$ holds, in the alternative (ii) at random. Thus Aristides is the representative of honesty; for, as the Athenians said, if there is any honest man it is Aristides. Assuming that we know the representative we can decide the question whether there is an honest man or whether all are dishonest by merely looking at *him*: if he is dishonest everybody is. In the realm of numbers we may even make the choice of the representative unique, in case (i) choosing $x = r$ as the *least* number for which $A(x)$ holds, and $r = 0$ in the opposite case (ii). Then r arises from A by a certain operator ϱ_x, $r = \varrho_x A(x)$, applicable to every imaginable property A. A propositional function may contain other variables $y, z, \cdots$ besides x.

Therefore it is necessary to attach an index x to ϱ, just as in integrating one must indicate with respect to which variable one integrates. ϱ_x eliminates the variable x; for instance $\varrho_x A(x, y)$ is a function of y alone. The word quantifier is in use for this sort of operator. Hence we write our axiom as follows:

$$A(b) \rightarrow A(\varrho_x A(x)). \tag{8}$$

It is immaterial whether we fix the representative in the unique manner described above; our specific rule would not fit anyhow unless x ranges over the numbers $0, 1, 2, \cdots$. Instead we imagine a quantifier ϱ_x of universal applicability which, as it were, selects the representative for us. Zermelo's axiom of choice is thus woven into the principle of the excluded middle. It is a bold step; but the bolder the better, as long as it can be shown that we keep within the bounds of consistency!

In the formalism, propositional functions are replaced by *formulas* the handling of which must be described without reference to their meaning. In general, variables $x, y, \cdots$ will occur among the symbols of a formula $\mathfrak{A}$. We say that the symbol ϱ_x *binds* the variable x in the formula $\mathfrak{A}$ which follows the symbol[12] and that x occurs *free* in a formula wherever it is not bound be a quantifier with index x. $x, y, \rightarrow, \varrho_x$ are symbols entering into the formulas; the German letters are no such symbols, but are used for communication. It is more natural to describe our critical axiom (8) as a rule for the formation of axioms. It says: take any formula $\mathfrak{A}$ in which only the variable x occurs free, and any formula $\mathfrak{b}$ without free variables, and by means of them build the formula

$$\mathfrak{A}(\mathfrak{b}) \rightarrow \mathfrak{A}(\varrho_x \mathfrak{A}). \tag{9}$$

Here $\mathfrak{A}(\mathfrak{b})$ stands for the formula derived from $\mathfrak{A}$ by putting in the entire formula $\mathfrak{b}$ for the variable x *wherever* x *occurs free.*

In this way formulas may be *obtained* as *axioms* according to certain rules. *Deduction* proceeds by the rule of syllogism: From two formulas $\mathfrak{a}$ and $\mathfrak{a} \rightarrow \mathfrak{b}$ *previously obtained*, in the second of which the first formula reappears at the left of the symbol $\rightarrow$, one *obtains* the formula $\mathfrak{b}$.

How does Hilbert propose to show that the game of deduction will never lead to the formula $0 \neq 0$? Here is the basic idea of his procedure. As long as

[12] If we wish the rule that ϱ_x binds x in all that comes after to be taken literally, we must write $\mathfrak{a} \rightarrow \mathfrak{b}$ in the form $\rightarrow \{ {\mathfrak{a} \atop \mathfrak{b}}$. The formulas will then look like genealogical trees.

one deals with "finite" formulas only, formulas from which the quantifiers ϱ_x, ϱ_y, $\cdots$ are absent, one can decide whether they are true or false by merely looking at them. With the entrance of ϱ such a descriptive valuation of formulas becomes impossible: evidence ceases to work. But a concretely given deduction is a sequence of formulas in which only a limited number of instances of the axiomatic rule (9) will turn up. Let us assume that the only quantifier which occurs is ϱ_x and wherever it occurs it is followed by the *same finite* formula $\mathfrak{A}$, so that the instances of (9) are of the form

$$\mathfrak{A}(\mathfrak{b}_1) \rightarrow \mathfrak{A}(\varrho_x\mathfrak{A}), \cdots, \mathfrak{A}(\mathfrak{b}_h) \rightarrow \mathfrak{A}(\varrho_x\mathfrak{A}). \tag{10}$$

Assume, moreover, $\mathfrak{b}_1, \cdots, \mathfrak{b}_h$ to be finite. We then carry out a *reduction*, replacing $\varrho_x\mathfrak{A}$ by a certain finite $\mathfrak{r}$ wherever it occurs as part of a formula in our sequence. In particular, the formulas (10) will change into

$$\mathfrak{A}(\mathfrak{b}_1) \rightarrow \mathfrak{A}(\mathfrak{r}), \cdots, \mathfrak{A}(\mathfrak{b}_h) \rightarrow \mathfrak{A}(\mathfrak{r}). \tag{11}$$

We now see how to choose $\mathfrak{r}$: if by examining the finite formulas $\mathfrak{A}(\mathfrak{b}_1), \cdots, \mathfrak{A}(\mathfrak{b}_h)$ one after the other, we find one that is true, say $\mathfrak{A}(\mathfrak{b}_3)$, then we take $\mathfrak{b}_3$ for $\mathfrak{r}$. If every one of them turns out to be false, we choose $\mathfrak{r}$ at random Then the h reduced formulas (11) are "true" and our hypothesis that the deduction leads to the false formula $0 \neq 0$ is carried *ad absurdum*. The salient point is that a concretely given deduction makes use of a limited number of explicitly exhibited individuals $\mathfrak{b}_1, \cdots, \mathfrak{b}_h$ only. If we make a wrong choice, for example, by choosing Alcibiades rather than Aristides as the representative of incorruptibility, our mistake will do no harm as long as the few people (out of the infinite Athenian crowd) with whom we actually deal are all corruptible.

A slightly more complicated case arises when we permit the $\mathfrak{b}_1, \cdots, \mathfrak{b}_h$ to contain ϱ_x, but always followed by the same $\mathfrak{A}$. Then we first make a *tentative* reduction replacing $\varrho_x\mathfrak{A}$ by the number 0, say. The formulas $\mathfrak{b}_1, \cdots, \mathfrak{b}_h$ are thus changed into reduced finite formulas $\mathfrak{b}_1^0, \cdots, \mathfrak{b}_h^0$ and (10) into

$$\mathfrak{A}(\mathfrak{b}_1^0) \rightarrow \mathfrak{A}(0), \cdots, \mathfrak{A}(\mathfrak{b}_h^0) \rightarrow \mathfrak{A}(0).$$

This reduction will do unless $\mathfrak{A}(0)$ is false and at the same time one of the $\mathfrak{A}(\mathfrak{b}_1^0), \cdots, \mathfrak{A}(\mathfrak{b}_h^0)$, say $\mathfrak{A}(\mathfrak{b}_3^0)$, is true. But then we have in $\mathfrak{b}_3^0$ a perfectly legitimate representative of $\mathfrak{A}$, and a second reduction which replaces $\varrho_x\mathfrak{A}$ by $\mathfrak{b}_3^0$ will work out all right.

However, this is only a modest beginning of the complications awaiting us. Quantifiers $\varrho_x, \varrho_y, \cdots$ with different variables and applied to different formulas will be piled one upon the other. We make a tentative reduction; it will go wrong in certain places and from that failure we learn how to correct it. But the corrected reduction will probably go wrong at other places. We seem to be driven around in a vicious circle, and the problem is to direct our consecutive corrections in such a manner as to obtain assurance that finally a reduction will result that makes good at all places in our given sequence of formulas. Nothing has contributed more to revealing the circle-like character of the usual transfinite arguments of mathematics than these attempts to make sure of consistency in spite of all circles.

The symbolism for the formalization of mathematics as well as the general layout and first steps of the proof of consistency are due to Hilbert himself. The program was further advanced by younger collaborators, P. Bernays, W. Ackermann, J. von Neumann. The last two proved the consistency of "arithmetics," of that part in which the dangerous axiom about the conversion of predicates into sets is not yet admitted. A gap remained which seemed harmless at the time, but already detailed plans were drawn up for the invasion of analysis. Then came a catastrophe: assuming that consistency is established, K. Gödel showed how to construct arithmetical propositions which are evidently true and yet not deducible within the formalism. His method applies to Hilbert's as well as any other not too limited formalism. Of the two fields, the field of formulas obtainable in Hilbert's formalism and the field of real propositions that are evidently true, neither contains the other (provided consistency of the formalism can be made evident). Obviously *completeness* of a formalism in the absolute sense in which Hilbert had envisaged it was now out of the question. When G. Gentzen later closed the gap in the consistency proof for arithmetics, which Gödel's discovery had revealed to be serious indeed, he succeeded in doing so only by substantially lowering Hilbert's standard of evidence.[13] The boundary line of what is intuitively trustworthy once more became vague. As all hands were needed to defend the homeland of arithmetics, the invasion of analysis never came off, to say nothing of general set theory.

This is where the problem now stands; no final solution is in sight. But whatever the future may bring, there is no doubt that Brouwer and Hilbert raised the problem of the foundations of mathematics to a new level. A

[13] G. Gentzen, Math. Ann. vol. 112 (1936) pp. 493—565.

return to the standpoint of Russell-Whitehead's *Principia Mathematica* is unthinkable.

Hilbert is the champion of axiomatics. The axiomatic attitude seemed to him one of universal significance, not only for mathematics, but for all sciences. His investigations in the field of physics are conceived in the axiomatic spirit. In his lectures he liked to illustrate the method by examples taken from biology, economics, and so on. The modern epistemological interpretation of science has been profoundly influenced by him. Sometimes when he praised the axiomatic method he seemed to imply that it was destined to obliterate completely the constructive or genetic method. I am certain that, at least in later life, this was not his true opinion. For whereas he deals with the primary mathematical objects by means of the axioms of his symbolic system, the formulas are constructed in the most explicit and finite manner. In recent times the axiomatic method has spread from the roots to all branches of the mathematical tree. Algebra, for one, is permeated from top to bottom by the axiomatic spirit. One may describe the role of axioms here as the subservient one of fixing the range of variables entering into the explicit constructions. But it would not be too difficult to retouch the picture so as to make the axioms appear as the masters. An impartial attitude will do justice to both sides; not a little of the attractiveness of modern mathematical research is due to a happy blending of axiomatic and genetic procedures.

Integral Equations

Between the two periods during which Hilbert's efforts were concentrated on the foundations, first of geometry, then of mathematics in general, there lie twenty long years devoted to analysis and physics.

In the winter of 1900–1901 the Swedish mathematician E. Holmgren reported in Hilbert's seminar on Fredholm's first publications on integral equations, and it seems that Hilbert caught fire at once. The subject has a long and tortuous history, beginning with Daniel Bernoulli. The mathematicians' efforts to solve the (mechanical, acoustical, optical, electromagnetical) problem of the oscillations of a continuum and the related boundary value problems of potential theory span a period of two centuries. Fourier's *Théorie analytique de la chaleur* (1822) is a landmark. H. A. Schwarz proved for the first time (1885) the existence of a proper oscillation in two and more dimensions by constructing the fundamental frequency of a membrane. The last decade of the nineteenth century saw Poincaré on his way to the development of powerful function-theoretic methods; C. Neumann and he came to grips with the harmonic boundary problem; Volterra studied that type of integral equations which now bears his name, and for linear equations with infinitely many unknowns Helge von Koch developed the infinite determinants. Most scientific discoveries are made when "their time is fulfilled"; sometimes, but seldom, a genius lifts the veil decades earlier than could have been expected. Fredholm's discovery has always seemed to me one that was long overdue when it came. What could be more natural than the idea that a set of linear equations connected with a discrete set of mass points gives way to an integral equation when one passes to the limit of a continuum? But the fact that in the simpler cases a differential rather than an integral equation results in the limit riveted the mathematicians' attention for two hundred years on differential equations!

It must be said, however, that the simplicity of Fredholm's results is due to the particular form of his equation, on which it was hard to hit without the guidance of the problems of mathematical physics to which he applied it:

$$x(s) - \int_0^1 K(s, t)x(t)dt = f(s) \qquad (0 \leqq s \leqq 1).$$

Indeed the linear operator which in the left member operates on the unknown function x producing a given f, $(E-K)x = f$, consists of two parts, the identity E and the integral operator K, which in a certain sense is weak compared to E. Fredholm proved that for this type of integral equation the two main facts about n linear equations with the same number n of unknowns hold: (1) The homogeneous equation $[f(s) = 0]$ has a finite number of linearly independent solutions $x(s) = \phi_1(s), \cdots, \phi_h(s)$, and the homogeneous equation with the transposed kernel $K'(s, t) = K(t, s)$ has the same number of solutions, $\psi_1(s), \cdots, \psi_h(s)$. (2) The nonhomogeneous equation is solvable if and only if the given f satisfies the h linear conditions

$$\int_0^1 f(s)\psi_i(s)ds = 0 \qquad (i = 1, \cdots, h).$$

Following an artifice used by Poincaré, Fredholm introduces a parameter λ replacing K by λK and obtains a solution in the form familiar from finite linear equations, namely as a quotient of two determinants of H. v. Koch's type, either of which is an entire function of the parameter λ.

Hilbert saw two things: (1) after having constructed Green's function K for a given region G and for the potential equation $\Delta u = 0$ by means of a

Fredholm equation on the boundary, the differential equation of the oscillating membrane $\Delta\phi + \lambda\phi = 0$ changes into a homogeneous integral equation

$$\phi(s) - \lambda \int_0^1 K(s, t)\phi(t)dt = 0$$

with the symmetric kernel K, $K(t, s) = K(s, t)$ (in which the parameter λ is no longer artificial but of the very essence of the problem); (2) the problem of ascertaining the "eigen values" λ and "eigen functions" $\phi(s)$ of this integral equation is the analogue for integrals of the transformation of a quadratic form of n variables onto principal axes. Hence the corresponding theorem for the quadratic integral form

$$\int_0^1 \int_0^1 K(s, t)x(s)x(t)dsdt \tag{12}$$

with an *arbitrary symmetric kernel* K must provide the general foundation for the theory of oscillations of a continuous medium. If others saw the same, Hilbert saw it at least that much more clearly that he bent all his energy on proving that proposition, and he succeeded by the same direct method which about 1730 Bernoulli had applied to the oscillations of a string: passage to the limit from the algebraic problem. In carrying out the limiting process he had to make use of the Koch-Fredholm determinant. He finds that there is a sequence of eigen values $\lambda_1, \lambda_2, \cdots$ tending to infinity, $\lambda_n \to \infty$ for $n \to \infty$, and an orthonormal set of corresponding eigen functions $\phi_n(s)$,

$$\phi_n(s) - \lambda_n \int_0^1 K(s, t)\phi_n(t)dt = 0,$$

$$\int_0^1 \phi_m(s)\phi_n(s)ds = \delta_{mn},$$

such that

$$\int_0^1 \int_0^1 K(s, t)x(s)x(t)dsdt = \sum \xi_n^2/\lambda_n,$$

ξ_n being the Fourier coefficient $\int_0^1 x(s)\phi_n(s)ds$. The theory implies that every function of the form

$$y(s) = \int_0^1 K(s, t)x(t)dt$$

may be expanded into a uniformly convergent Fourier series in terms of the eigen functions ϕ_n,

$$y(s) = \sum \eta_n \phi_n(s), \qquad \eta_n = \int_0^1 y(s)\phi_n(s)ds.$$

Hilbert's passage to the limit is laborious. Soon afterwards E. Schmidt in a Göttingen thesis found a simpler and more constructive proof for these results by adapting H. A. Schwarz's method invented twenty years before to the needs of integral equations.

From finite forms the road leads either to integrals or to infinite series. Therefore Hilbert considered the same problem of orthogonal transformation of a given quadratic form

$$\sum K_{mn} x_m x_n \tag{13}$$

into a form of the special type

$$\varkappa_1 \xi_1^2 + \varkappa_2 \xi_2^2 + \cdots \qquad (\varkappa_n = 1/\lambda_n \to 0) \tag{14}$$

also for infinitely many (real) variables $(x_1, x_2, \cdots)$ or vectors x in a space of a denumerable infinity of dimensions. Only such vectors are admitted as have a finite length $|x|$,

$$|x|^2 = x_1^2 + x_2^2 + \cdots;$$

they constitute what we now call the Hilbert space. The advantage of Hilbert space over the "space" of all continuous functions $x(s)$ lies in a certain property of completeness, and due to this property one can establish "complete continuity" as the necessary and sufficient condition for the transformability of a given quadratic form K (13) into (14) by following an argument well known in the algebraic case: one determines $\varkappa_1, \varkappa_2, \cdots$ as the consecutive maxima of K on the "sphere" $|x|^2 = 1$.

As suggested by the theorem concerning a quadratic integral form, the link between the space of functions $x(s)$ and the Hilbert space of vectors $(x_1, x_2, \cdots)$ is provided by an arbitrary *complete* orthonormal system $u_1(s)$, $u_2(s), \cdots$ and expressed by the equations

$$x_n = \int^1 x(s)u_n(s)ds.$$

Bessel's inequality states that the square sum of the Fourier coefficients x_n is less than or equal to the square integral of $x(s)$. The relation of completeness, first introduced by A. Hurwitz and studied in detail by W. Stekloff, requires that in this inequality the *equality* sign prevail. Thus the theorem on quadratic forms of infinitely many variables at once gives the corresponding results about the eigen values and eigen functions of symmetric kernels $K(s, t)$ — or would do so if one could count on the uniform convergence of $\sum x_n u_n(s)$ for any given vector $(x_1, x_2, \cdots)$ in Hilbert space. In the special case of an eigen vector of that quadratic form (13) which corresponds to the integral form (12),

$$x_n = \lambda \sum_m K_{nm} x_m ,$$

Hilbert settles this point by forming the uniformly convergent series

$$\lambda \sum_m x_m \int_0^1 K(s, t) u_m(t) dt$$

which indeed yields a continuous function $\phi(s)$ with the nth Fourier coefficient

$$\lambda \sum K_{nm} x_m = x_n,$$

and thus obtains the eigen function of $K(s, t)$ for the eigen value λ. Soon afterwards, under the stimulus of Hilbert's investigations, E. Fischer and F. Riesz proved their well known theorem that the space of all functions $x(s)$ the square of which has a finite Lebesgue integral enjoys the same property of completeness as Hilbert space, and hence one is mapped isomorphically upon the other in a one-to-one fashion by means of a complete orthonormal system $u_n(s)$. I mention these details because the historic order of events may have fallen into oblivion with many of our younger mathematicians, for whom Hilbert space has assumed that abstract connotation which no longer distinguishes between the two realizations, the square integrable functions $x(s)$ and the square summable sequences $(x_1, x_2, \cdots)$. I think Hilbert was wise to keep within the bounds of continuous functions when there was no actual need for introducing Lebesgue's general concepts.

Perhaps Hilbert's greatest accomplishment in the field of integral equations is his extension of the theory of spectral decomposition from the completely continuous to the so-called *bounded* quadratic forms. He finds that then the point spectrum will in general have condensation points and a continuous spectrum will appear beside the point spectrum. Again he proceeds by directly carrying out the transition to the limit, letting the number of variables $x_1, x_2, \cdots$ increase *ad infinitum*. Again, not long afterwards, simpler proofs for his results were found.

While thus advancing the boundaries of the general theory, he did not lose sight of the ordinary and partial differential equations from which it had sprung. Simultaneously with the young Italian mathematician Eugenio Elia Levi he developed the parametrix method as a bridge between differential and integral equations. For a given elliptic differential operator Δ^* of the second order, the parametrix $K(s, t)$ is a sort of qualitative approximation of Green's function, depending like the latter on an argument point s and a parameter point t. It is supposed to possess the right kind of singularity for $s = t$ so that the nonhomogeneous equation $\Delta^* u = f$ for

$$u = K\varrho, \qquad u(s) = \int K(s, t)\varrho(t)dt$$

gives rise to the integral equation $\varrho + L\varrho = f$ for the density ϱ, with a kernel $L(s, t) = \Delta_s^* K(s, t)$ regular enough at $s = t$ for Fredholm's theory to be applicable. It is important to give up the assumption that K satisfies the equation $\Delta^* K = 0$, because in general a fundamental solution will not be known for the given differential operator Δ^*. In order not to be bothered by boundary conditions, Hilbert assumes the domain of integration to be a *compact* manifold, like the surface of a sphere, and finds that the method works if the parametrix, besides having the right kind of singularity, is symmetric with respect to argument and parameter.

What has been said should be enough to make clear that in the terrain of analysis a rich vein of gold had been struck, comparatively easy to exploit and not soon to be exhausted. The linear equations of infinitely many unknowns had to be investigated further (E. Schmidt, F. Riesz, O. Toeplitz, E. Hellinger, and others); the continuous spectrum and its appearance in integral equations with "singular" kernels awaited closer analysis (E. Hellinger, T. Carleman); ordinary differential equations, with regular or singular boundaries, of second or of higher order, received their due share of attention (A. Kneser, E. Hilb, G. D. Birkhoff, M. Bôcher, J. D. Tamarkin, and many others).[14] It became possible to develop such asymptotic laws for the distribution of eigen values as were required by the thermodynamics of radiation (H. Weyl, R. Courant). Expansions in terms of orthogonal func-

[14] For later literature and systems of differential equations see Axel Schur, Math. Ann. vol. 82 (1921) pp. 213—239; G. A. Bliss, Trans. Amer. Math. Soc. vol. 28 (1926) pp. 561—584; W. T. Reid, ibid vol. 44 (1938) pp. 508—521.

tions were studied independently of their origin in differential or integral equations. New light fell upon Stieltjes's continued factions and the problem of momentum. The most ambitious began to attack nonlinear integral equations. A large international school of young mathematicians gathered around Hilbert and integral equations became the fashion of the day, not only in Germany, but also in France where great masters like E. Picard and Goursat paid their tributes, in Italy and on this side of the Atlantic. Many good papers were written, and many mediocre ones. But the total effect was an appreciable change in the aspect of analysis.

Remarkable are the applications of integral equations outside the field for which they were invented. Among them I mention the following three: (1) Riemann's problem of determining n analytic functions $f_1(z), \cdots, f_n(z)$, regular except at a finite number of points, which by analytic continuation around these points suffer preassigned linear transformations. The problem was solved by Hilbert himself, and subsequently in a simpler and more complete form by J. Plemelj. (A very special case of it is the existence of algebraic functions on a Riemann surface if that surface is given as a covering surface of the complex z-plane.) Investigations by G. D. Birkhoff on matrices of analytic functions lie in the same line. (2) Proof for the completeness of the irreducible representations of a compact continuous group. This is an indispensable tool for the approach to the general theory of invariants by means of Adolf Hurwitz's integration method, and with its refinements and generalizations plays an important role in modern group-theoretic research, including H. Bohr's theory of almost periodic functions.[15] Contact is thus made with Hilbert's old friend, the theory of invariants. (3) Quite recently Hilbert's parametrix method has served to establish the central existence theorem in W. V. D. Hodge's theory of harmonic integrals in compact Riemannian spaces.[16]

The story would be dramatic enough had it ended here. But then a sort of miracle happened: the spectrum theory in Hilbert space was discovered to be the adequate mathematical instrument of the new quantum physics inaugurated by Heisenberg and Schrödinger in 1925. This latter impulse led to a reexamination of the entire complex of problems with refined means (J. von Neumann, A. Wintner, M. H. Stone, K. Friedrichs). As J. von

[15] H. Weyl and F. Peter, Math. Ann. vol. 97 (1927) pp. 737—755. A. Haar, Ann. of Math. vol. 34 (1933) pp. 147—169. J. von Neumann, Trans. Amer. Math. Soc. vol. 36 (1934) pp. 445—492. Cf. also L. Pontrjagin, *Topological groups*, Princeton 1939.

[16] W. V. D. Hodge, *The theory and applications of harmonic integrals*, Cambridge, 1941. H. Weyl, Ann. of Math. vol. 44 (1943) pp. 1—6.

Neumann was Hilbert's collaborator toward the close of that epoch when his interest was divided between quantum physics and foundations, the historic continuity with Hilbert's own scientific activities is unbroken, even for this later phase. What has become of the theory of abstract spaces and their linear operators in our times lies beyond the bounds of this report.

A picture of Hilbert's "analytic" period would be incomplete without mentioning a second motif, *calculus of variations*, which crossed the dominating one of integral equations. The "theorem of independence" with which he concludes his Paris survey of mathematical problems (1900) is an important contribution to the formal apparatus of that calculus. But of much greater consequence was his audacious direct assault on the functional maxima and minima problems. The whole finely wrought machinery of the calculus of variations is here consciously set aside. He proposes instead to construct the minimizing function as the limit of a sequence of functions for which the value of the integral under investigation tends to its minimum value. The classical example is Dirichlet's integral in a two-dimensional region G,

$$D[u] = \iint_G \left\{ \left(\frac{\partial u}{\partial x}\right)^2 + \left(\frac{\partial u}{\partial y}\right)^2 \right\} dx dy.$$

Admitted are all functions u with continuous derivatives which have given boundary values. d being the lower limit of $D[u]$ for admissible u, one can ascertain a sequence of admissible functions u_n such that $D[u_n] \to d$ with $n \to \infty$. One cannot expect the u_n themselves to converge; rather they have to be prepared for convergence by the smoothing process of integration. As the limit function will be harmonic and the value of the harmonic function at any point P equals its mean value over any circle K around P, it seems best to replace $u_n(P)$ by its mean value in K, with the expectation that this mean value will converge toward a number $u(P)$ which is independent of the circle and in its dependence on P solves the minimum problem. Besides integration Hilbert uses the process of sifting a suitable subsequence from the u_n before passing to the limit. Owing to the simple inequality

$$\{D[u_m - u_n]\}^{1/2} \leqq \{D[u_m] - d\}^{1/2} + \{D[u_m] - d\}^{1/2}$$

discovered by S. Zaremba this second step is unnecessary.

Hilbert's method is even better suited for problems in which the boundary does not figure so prominently as in the boundary value problems. By a slight modification one is able to include point singularities, and Hilbert thus solved the fundamental problem for flows on Riemann surfaces, providing thereby the necessary foundation for Riemann's own approach to the theory of Abelian integrals, and he further showed that Poincaré's and Koebe's fundamental theorems on uniformization could be established in the same way. We should be much better off in number theory if methods were known which are as powerful for the construction of relative Abelian and Galois fields over given algebraic number fields as the Riemann-Hilbert transcendental method proves to be for the analogous problems in the fields of algebraic functions! Its wide application in the theory of conformal mapping and of minimal surfaces is revealed by the work of the man who was Hilbert's closest collaborator in the direction of mathematical affairs at Göttingen for many years, Richard Courant.[17] Of a more indirect character, but of considerable vigor, is the influence of Hilbert's ideas upon the whole trend of the modern development of the calculus of variations; in Europe Carathéodory, Lebesgue, Tonelli could be mentioned among others, in this country the chain reaches from O. Bolza's early to M. Morse's most recent work.

Physics

Already before Minkowski's death in 1909, Hilbert had begun a systematic study of theoretical physics, in close collaboration with his friend, who had always kept in touch with the neighboring science. Minkowski's work on relativity theory was the first fruit of these joint studies. Hilbert continued them through the years, and between 1910 and 1930 often lectured and conducted seminars on topics of physics. He greatly enjoyed this widening of his horizon and his contact with physicists, whom he could meet on their own ground. The harvest however can hardly be compared with his achievements in pure mathematics. The maze of experimental facts which the physicist has to take into account is too manifold, their expansion too fast, and their aspect and relative weight too changeable for the axiomatic method to find a firm enough foothold, except in the thoroughly consolidated parts of our physical knowledge. Men like Einstein or Niels Bohr grope their way in the dark toward their conceptions of general relativity or atomic structure by another type of experience and imagination than those of the

---

[17] A book by Courant on the Dirichlet principle is in preparation. [Editor's note: R. Courant, *Dirichlet's Principle, conformal mapping, and minimal surfaces*, Interscience Publishers Inc., New York, 1950.]

mathematician, although no doubt mathematics is an essential ingredient. Thus Hilbert's vast plans in physics never matured.

But his application of integral equations to kinetic gas theory and to the elementary theory of radiation were notable contributions. In particular, his asymptotic solution of Maxwell-Boltzmann's fundamental equation in kinetic gas theory, which is an integral equation of the second order, clearly separated the two layers of phenomenological physical laws to which the theory leads; it has been carried out in more detail by the physicists and applied to several concrete problems. In his investigations on general relativity Hilbert combined Einstein's theory of gravitation with G. Mie's program of pure field physics. For the development of the theory of general relativity at that stage, Einstein's more sober procedure, which did not couple the theory with Mie's highly speculative program, proved the more fertile. Hilbert's endeavors must be looked upon as a forerunner of a unified field theory of gravitation and electromagnetism. However, there was still much too much arbitrariness involved in Hilbert's Hamiltonian function; subsequent attempts (by Weyl, Eddington, Einstein himself, and others) aimed to reduce it. Hopes in the Hilbert circle ran high at that time; the dream of a universal law accounting both for the structure of the cosmos as a whole, and of all the atomic nuclei, seemed near fulfillment. But the problem of a unified field theory stands to this day as an unsolved problem; it is almost certain that a satisfactory solution will have to include the material waves (the Schrödinger-Dirac Ψ for the electron, and similar field quantities for the other nuclear particles) besides gravitation and electromagnetism, and that its mathematical frame will not be a simple enlargement of that of Einstein's now classical theory of gravitation.

Hilbert was not only a great scholar, but also a great teacher. Witnesses are his many pupils and assistants, whom he taught the handicraft of mathematical research by letting them share in his own work and its overflow, and then his lectures, the notes of many of which have found their way from Göttingen into public and private mathematical libraries. They covered an extremely wide range. The book he published with S. Cohn-Vossen on *Anschauliche Geometrie* is an outgrowth of his teaching activities. Going over the impressive list attached to his Collected Papers (vol. 3, p. 430) one is struck by the considerable number of courses on general topics like "Knowledge and Thinking," "On the Infinite," "Nature and Mathematics." On the whole, his lectures were a faithful reflection of his spirit; direct, intense; how could they fail to be inspiring?

Index of Names

Page numbers in *italics* refer to the illustrations